PRAISE FOR *BURN*

"We're in a climate emergency, and we need to be using an awful lot of different approaches — here's one that definitely deserves to be explored in full."
— BILL MCKIBBEN, author of *Falter*

"*Burn* advances the discussion from fantasies of biochar-based agriculture to normative proposals for many ways the material could theoretically be used as an environmentally attractive, economically competitive resource in many sectors of society. The book opens new avenues of thought, and it will be a valuable reference in the coming decade in helping us to assess the inevitable cascade of ever bigger, riskier, costlier, and zanier proposals for carbon withdrawal."
— DENNIS MEADOWS, 2018 laureate, The Earth Hall of Fame Kyoto

"For anyone interested in solutions to climate change, this book is absolutely essential reading. It represents the latest, most innovative thinking and experimentation on removing carbon from the atmosphere. What's delightfully startling is the authors' detailed, example-laden argument that we can use carbon to regenerate landscapes while also producing an astounding array of products — from concrete to plastics to batteries to paper — that function better by incorporating the universe's most versatile element. Written in a clear, entertaining style, *Burn* is an incendiary contribution."
— RICHARD HEINBERG, senior fellow, Post Carbon Institute

"I cannot recommend this book highly enough for going deep into the science of a potentially revolutionary technology that could be capable of stopping dangerous climate change in its tracks. For anyone who wants to know how societies can transform the very fabric of how we run our industries so that we protect and enhance our environment, not destroy it — while contributing to thriving economies — this is literally the manual. It is, in short, a window into the future we could build together. So read it, and start building."
— DR. NAFEEZ AHMED, system shift columnist, Motherboard;
editor-in-chief, INSURGE Intelligence;
research fellow, The Schumacher Institute

"What if we could make carbon our ally, instead of our enemy, in preserving this planet? This deeply detailed book is about far more than the ancient, carbon-fixing Amazonian soil technology called *terra preta*. Practically everything humans do, *Burn* shows, could reimburse the Earth for the carbon we've exhumed, leaving civilization far cleaner and healthier — and with a chance for a future."

— ALAN WEISMAN, author of *Countdown*,
The World Without Us, and *Gaviotas*

"Carbon, the most promiscuous of elements, can be our ruination or by better management, our salvation. *Burn* is a clear, accessible, and luminescent blueprint for the latter. It really is a must-read."

— DAVID ORR, author of *Dangerous Years*

"Carbon is the element that likes to hold hands and collaborate. We can learn a lot from carbon if we stop demonizing it. *Burn* does an exceptional job telling the vital story of how carbon can address the interconnected crises in waste, energy, food, soil, water, and, most pressingly, climate. This book plays a critical role in educating us to reorient with carbon math, reimagine the role of carbon cascades, and redesign the carbon cycle."

— AMANDA JOY RAVENHILL, executive director,
Buckminster Fuller Institute; cofounder, Project Drawdown

"This book is a big deal. It argues persuasively that carbon has been vilified for far too long. Biochar, a hard, crystal-like form of carbon, can reanimate tired soils and help to mop up vast amounts of CO_2 from the atmosphere. The authors speak as seasoned scientists as well as practitioners, and their arsenal of arguments offers more than a glimpse of hope in a world threatened with climate doom. If there is a way out, here is a bunch of keys to the door at the end of the tunnel."

— HERBERT GIRARDET, cofounder, World Future Council;
executive council member, Club of Rome

"Brilliant in its range and depth, *Burn* offers an integrated approach to addressing climate change and biodiversity loss and provides potential solutions for tackling the full range of activities that negatively impact our climate. It is a groundbreaking sequel to *The Paris Agreement* and gives hope to a world currently facing a multiplicity of interlinked crises."

— FEARGAL DUFF, environmental activist

BURN

ALSO BY ALBERT BATES

Shutdown! Nuclear Power on Trial

Climate in Crisis:
The Greenhouse Effect and What We Can Do

The Post-Petroleum Survival Guide and Cookbook:
Recipes for Changing Times

The Biochar Solution:
Carbon Farming and Climate Change

The Paris Agreement:
The Best Chance We Have to Save
the One Planet We've Got

ALSO BY KATHLEEN DRAPER

Terra Preta:
How the World's Most Fertile Soil Can Help
Reverse Climate Change and Reduce World Hunger

BURN

Using Fire to Cool the Earth

Albert Bates and **Kathleen Draper**

CHELSEA GREEN PUBLISHING
White River Junction, Vermont
London, UK

Project Manager: Alexander Bullett
Developmental Editor: Brianne Goodspeed
Copy Editor: Deborah Heimann
Proofreader: Laura Jorstad
Indexer: Linda Hallinger
Designer: Melissa Jacobson

Printed in Canada.
First printing February, 2019.
10 9 8 7 6 5 4 3 2 1   19 20 21 22 23

Our Commitment to Green Publishing
Chelsea Green sees publishing as a tool for cultural change and ecological stewardship. We strive to
align our book manufacturing practices with our editorial mission and to reduce the impact of our
business enterprise in the environment. We print our books and catalogs on chlorine-free recycled
paper, using vegetable-based inks whenever possible. This book may cost slightly more because it was
printed on paper that contains recycled fiber, and we hope you'll agree that it's worth it. Chelsea Green
is a member of the Green Press Initiative (www.greenpressinitiative.org), a nonprofit coalition of
publishers, manufacturers, and authors working to protect the world's endangered forests and conserve
natural resources. *Burn* was printed on paper supplied by Marquis that contains 100% postconsumer
recycled fiber.

Library of Congress Cataloging-in-Publication Data
Names: Bates, Albert K., 1947- author. | Draper, Kathleen, author.
Title: Burn : using fire to cool the earth / Albert Bates and Kathleen Draper.
Description: White River Junction, Vermont : Chelsea Green Publishing, [2019] | Includes index.
Identifiers: LCCN 2018047434| ISBN 9781603587839 (hardcover) | ISBN 9781603587846 (ebook)
Subjects: LCSH: Climate change mitigation--Popular works. | Carbon dioxide
  mitigation--Popular works. | Biomass energy--Popular works. | Climatic changes--Social
  aspects--Popular works. | Global warming--Social aspects--Popular works.
Classification: LCC TD171.75 .B38 2019 | DDC 363.738/746--dc23
LC record available at https://lccn.loc.gov/2018047434

Chelsea Green Publishing
85 North Main Street, Suite 120
White River Junction, VT 05001
(802) 295-6300
www.chelseagreen.com

RECYCLED
Paper made from
recycled material
FSC® C103567

To Hans-Peter Schmidt and Gunter Pauli.
Where others see only separation, they perceive connection.

CONTENTS

Introduction

We tend to conceive of evolution as a process that occurs gradually and steadily over millions of years, but actually evolution happens in fits and starts — very slowly for long periods, then in sudden spurts of rapid change. Often the trigger is a particular event or convergence of events that shakes up the order of things. Within a very short time afterward, new life-forms emerge, ecotones form, and long-established orders realign. Evolutionary biologist Stephen J. Gould called this process "punctuated equilibrium."

History, sociology, and anthropology show us that cultural evolution proceeds in much the same way. Civilizations are living entities with regular cycles of birth, growth, and death. They may evolve and grow for as little as a century or two, as the Inca empire did, or over the course of thousands of years, as with India and China. When a civilization begins, it is a child. It tries new things and adopts behaviors it admires. As it matures, its social norms become more rigid, embedded, and brittle. It may lose some of its ability to respond to change or adapt in new ways. Each generation is taught to accept "the way things are" without questioning. This phase often ends in corruption, decay, and decline.

Many of us can sense the next punctuation coming. Perhaps it has already begun. The starting point for the current phase might have been three centuries ago. At that moment humans had only just discovered how to harness coal to make steam but had yet to employ the far greater energy density of oil and gas, never mind nuclear fission. The mere addition of coal to the human energy portfolio could well have been enough to augur the end of the global civilization we know today. Civilization itself, in the words of ecologist Tim Garrett, has become a heat engine.[1]

Coal from the Fushun mine in northeastern China was used to smelt copper as early as 1000 BCE but it was the advent of James Watt's steam engine in the eighteenth century that gave fossil energy traction, literally. In perfect parallel, expansion of the human population tracked expansion of the supply of available energy, railroads, and factories. Thomas Malthus, running the mathematical equations for population, and the Swedish Nobel Prize winner Svante Arrhenius, doing the same for climate change roughly a century later, accurately predicted the outcome once humanity was swept up in the enchantment of seemingly unlimited energy.

As we progressed in our ability to harness energy, we moved from a nearly stable world population, fluctuating little over the course of thousands of years, to a steady growth rate of 30 percent every twenty years. In 1925, we'd added one billion more people to the planet and 20 parts per million (ppm) of carbon dioxide (CO_2) to the atmosphere over the previous seventy-five years (from 1850). We now add one billion people to the planet and 25 to 30 ppm of CO_2 to the atmosphere every twelve years.

Historian William Catton labeled modern humans *Homo colossus* — those among us living in industrial countries and consuming massive amounts of fossil fuels to motivate and control machines that do orders of magnitude more work than humans or animals could do alone. Rolling along like a quiet juggernaut, *Homo colossus* is replacing *Homo sapiens* across the face of the Earth. While *Homo sapiens*, with a stable population under one billion, might have had a reasonable chance of being around for another two or three million years, *Homo colossus* hasn't a prayer.

A reckoning awaits. When, exactly, that reckoning might occur is difficult to predict. It could occur suddenly, as the deceptive debt instruments engineered to cover the real life-support deficit can no longer be serviced. It could occur slowly, as we continue squeezing out the last tons of brown coal, barrels of tarry shale oil, and cubic meters of unconventional gas, using ever-advancing technologies to find, refine, and burn them as quickly as possible, ignoring the horrific climate consequences we are locking in.

Evolutionary biologist Bruce H. Lipton says there are three questions that form the base paradigm of civilizations.[2] If the old answers are wrong, or become wrong over time, new answers are required. Civilizations that stay nimble enough to adopt the new answers begin a new chapter of life. Those that don't, disappear. The three questions are:

1. How did we get here?
2. Why are we here?
3. How can we make the best of it?

Introduction

The answer to the first question is a wild story no matter how you approach it. You could say we are here because billions of years ago astronomical collisions occurred as objects moving out from the Big Bang ricocheted like billiard balls and in an extraordinary chance occurrence one of those collisions produced an elliptical orbit in the third planet from a star, an orbiting moon just the right distance from that planet to pull tides, a spin that secured climate gradients between the poles and equator, an eccentric tilt of the axis that permitted annual seasons and the ebb and flow of photosynthesis. In these extraordinarily auspicious circumstances of birth we were also given the rarest gift: the presence of surface water, arriving like the rupture of the amniotic sac at the start of labor.

The collision that struck off Earth's moon enveloped the young Earth in a hot metallic vapor: 230°C (446°F). Over the course of a few thousand years, that vapor condensed, perspiring water, and leaving behind a sweltering CO_2 atmosphere. Liquid oceans formed despite the temperature because of the pressure of the heavy atmosphere. Gradually, subduction by plate tectonics and absorption by ocean water removed most CO_2 from the atmosphere, cooling the world and yielding saline oceans and a benign atmosphere of oxygen, hydrogen, and nitrogen — and the perfect conditions for life to arise.

Or this all may just be a dream that Vishnu is having.

Why we are here is anyone's guess. Perhaps it was so that biological organisms could set foot on other planets, by fugitive bacteria hitching a ride with the Mars rover *Curiosity*. Or perhaps we are nature's way of generating another mass extinction event in order to clear the stage for the next evolutionary drama.

If a civilization answers the third question in a way that ignores the energy and resource flows and storages of the planet — "get more stuff," "watch out for number one," or "this world doesn't matter, it is the next we want to get into" — destruction will be our destiny. If, on the other hand, a civilization says, "maintain harmony," "don't anger the gods," or "live lightly and plan for the future," restoration and continuation can be our happy outcome.

Right now the majority of people in the world cling to self-destructive ways. They are set in old patterns and don't realize how fragile and brittle those are. A growing minority see better ways and are putting together the building blocks for a new phase. We had the pleasure of meeting many of those pioneers in the process of writing this book.

This is a book about carbon. It is also about climate change, and how that really changes everything now. This book is not intended to persuade anyone to a particular point of view. It is addressed to those who have grasped the hard truth that human ecology will need to change if we are to remain. This is for those of us who are looking for a way forward that is possible, viable, and achievable. No greenwash. Practical. Doable. But the solutions are complicated, multidimensional, cross-cutting, and therefore prone to overwhelm. Our wish is to make this subject accessible.

Through many of the chapters that follow we often linger on a particular form of carbon: the solid, chemically recalcitrant form called biochar, obtained from the carbonization of biomass. Rediscovered in the latter half of the twentieth century in the famous *terra preta do indios* — the Amazonian dark earths — biochar has become the subject of increasingly intensive scientific research, and for good reason. It possesses a variety of remarkable attributes amenable to many different services and products, nearly all of which ultimately help rebalance atmospheric carbon and improve ecosystem health.

According to the International Biochar Initiative, "biochar can be distinguished from charcoal — used mainly as a fuel — in that a primary application is use as a soil amendment with the intention to improve soil functions and to reduce emissions from biomass that would otherwise naturally degrade to greenhouse gases."[3]

Indeed, when people discuss biochar, they often focus almost exclusively on agricultural uses. This is admittedly the focus of *The Biochar Solution* (published by Albert in 2010) and *Terra Preta* (published by Kathleen with Ute Scheub, Haiko Pieplow, and Hans-Peter Schmidt in 2016). Drawing on the published science at those times, our best estimate for biochar's potential drawdown effect worldwide — the amount of carbon that could be removed from the atmosphere and oceans — was about 1 gigaton of carbon per year (GtC/yr). Current annual anthropogenic emissions are more than 9.5 GtC/yr.[4] After adjusting for the difference between carbon and carbon dioxide — 1 ton of carbon equals 3.67 tons of carbon dioxide — that's 35–40 gigatons of carbon dioxide and its equivalents in other greenhouse gases (CO_2-e) every year. The shortfall is rather glaring.

Introduction

That sequestration potential suggested in 2010 is only 10 percent of the annual human carbon footprint. Carbon farming in all its permutations, as necessary and useful as it is, is unlikely to be sufficient to reverse climate change on its own.

There is good news, however. Really good news. Evidence is rapidly mounting that the potential of harnessing carbon to reverse climate change extends far beyond agriculture. That potential is what this book is all about. We seek to share the growing number of ways that carbon can be used above the ground to improve health, rebuild infrastructure, provide or boost renewable-energy production, rebalance atmosphere and ocean equilibrium, and offer a host of other benefits. Practitioners from around the world are demonstrating that carbonizing biomass can help urban and rural areas become more sustainable and regenerative by reducing waste, restoring ecosystems, closing nutrient cycles, and slashing emissions.

In writing this book, our travels took us, together or separately, to rice paddies in Qinfeng (located in Liuhe District of China) where we watched rice straw being processed into biochar and wood vinegar; to a ridge looking down an agro-forested valley on the island of Hispañola toward the beach where Columbus made landfall in 1492; to mountain villages in Costa Rica and Nepal where biochar recipes were being tested as soaps and toothpastes or in cement tiles to repair roofs damaged by hurricanes and earthquakes; to a willow plantation supplying heat for Gurteen College in County Tipperary, Ireland; to demonstrations of various homemade pyrolyzing devices at the New England Small Farm Institute; to crop research plots in California, Iowa, and Oregon. We made biochar in open pits at permaculture gatherings in England, Estonia, and India; demonstrated camp stoves that charged cell phones to billionaire investors and hedge fund managers in Tulum, Paris, Marrakech, and Bonn; and trekked into remote jungles in Mexico and Belize to scope out microenterprise hubs emerging around experimental biorefineries.

We had frank discussions with Anote Tong, former president of Kiribati; Patricia Scotland, secretary general of the Commonwealth of Nations; Thomas J. F. Goreau of the Coral Reef Alliance; Christiana Figueres of Mission 2020; and John Dennis Liu of the Ecosystem Regeneration Camps. We spoke with authors David Yarrow, Paul Hawken, Daniel Christian Wahl, and Tim Flannery and interviewed scientists Bronson Griscom, Charles A. S.

Hall, Johannes Lehmann, Annette Cowie, Dennis Meadows, Guy McPherson, Kevin Anderson, and Janine Benyus. Kathleen led biochar study tours to learn from projects in Austria, Sweden, and Nepal. Albert traveled to India, China, and Russia under the auspices of the Global Ecovillage Network. We attended webinars and conferences with subjects as diverse as gardening, biomass energy, and blockchain. Everywhere we went and everyone with whom we spoke lent a sense of urgency to the publication of this book.

As we'll cover in part 1, carbon has nearly unmatched versatility, largely untapped at this point in history. By understanding carbon, we can reimagine our relationship to it. Instead of constantly maligning it, we should be learning to put carbon to better and better uses. We'll explore the natural carbon cycle and how we have allowed ourselves to get disastrously out of balance with it. Once we understand carbon — and how, in particular, we can go from squandering carbon to banking it in a virtuous cycle of improvement we refer to as "carbon cascades" — we can begin to see the massive opportunity it presents, rather than only the threat of planetary proportions we've all been so focused on.

In part 2, we examine how the built environment, from pavements and bridges to airports and skyscrapers, can benefit from adding more carbon. Those who get into carbon cascades early may find wealth opportunities on a par with the great railroads, steel mills, and steamships of the nineteenth century. It's not the Industrial Revolution, but it's a carbon copy. We also look at how carbon can transform major social services and infrastructure, such as the provision of clean(er) water, radical improvements to the food supply, and waste management that doesn't rely on dumping trash on low-income communities. We show how by retasking carbon we can restore coral reefs, turn back deserts, and reforest damaged and denuded landscapes.

Part 3 gets personal. We look at ways carbon can be used to improve our lives, including the products we surround ourselves with and use on a daily basis. While scaling back consumer patterns is crucial to leaving a livable planet to future generations, we also take a practical look at how, with the right approach, we can have our cars and drive them, too. We cover cutting-edge technologies for refrigeration, batteries, pet care, housewares — even health and wellness. Carbon-rich products will help us transition to the next economic paradigm. We can reverse climate change by redesigning human ecology. We call this our displacement strategy.

Finally, in part 4, we'll address the problem of change itself. What are the impediments? What holds us back? Better still, what are the catalysts? What could transform our governmental, social, and commercial structures quickly enough to avert climate scientists' most dire predictions? What is our vision for the future? What world do we want to live in and leave to our children, and how to we get there?

If we are going to make investments, we'll need to know what *is* the potential carbon cascades can offer to rebalance the chemistry of Earth's atmosphere and oceans. In the "Carbon Math" tables throughout the book, we've attempted to quantify the drawdown from feedstocks that could be transformed from cyclical, photosynthetic, biologically active carbon to withheld, mineralized, stored carbon.

We also address a concern that often arises in attempts to quantify the drawdown potential of biochar — where can we put it? We've used these tables to catalog many major industrial displacement opportunities, moving beyond agriculture and looking at everything from tires and toothpaste to roads and bridges. We hope these beginnings will stimulate research to improve upon our tentative estimates. What these numbers provide is the first glimpse of one possible future for the twenty-first century: an eco-civilization retooled to reverse climate change.

———————

We have a choice before us now, individually and collectively. Civilizations undergo transformations. We can leave behind the old one that is poorly adapted, and design and build a better society. This book is part of that visioning process. The destructive civilization of the past few centuries was founded on plundering and profiting from prehistoric carbon. The new economy will be carbon-centric, too, but the focus will be on continuous cycling — and a virtuous spiral of improvement.

As the planet teeters on a climate precipice and the global economy is running at full speed within a fossil-carbon-induced haze, many people see no viable solutions to looming interconnected disasters. Those few among us who have glimpsed the possibility for a new carbon economy may seem naive. But these are neither moonshots nor science fiction. They are economically viable reconceptions for our global industrial model. Some solutions are already being field-tested while others have yet to leave the laboratory. What this means is that it is an exciting time to be carbon beings in a carbon world, learning how to grow and prosper with the natural cycles of our planet rather than constantly fighting against them. While we are quick to agree with the world's best science that the challenge is daunting and the hour is late, we believe that hidden within the problem lie the seeds of a solution. The central seed is the element carbon, and this book describes how to transform it from nemesis to ally: by learning to see the unprecedented opportunity embedded within the threat.

PART I

Carbon Change: From Nemesis to Ally

Everything the Power of the World does is done in a circle. The sky is round, and I have heard that the earth is round like a ball, and so are all the stars. The wind, in its greatest power whirls. Birds make their nest in circles, for theirs is the same religion as ours. The sun comes forth and goes down again in a circle. The moon does the same and both are round. Even the seasons form a great circle in their changing, and always come back again to where they were. The life of a man is a circle from childhood to childhood, and so it is in everything where power moves. Our tepees were round like the nests of birds, and these were always set in a circle, the nation's hoop.

— BLACK ELK, *Oglala Lakota* [1]

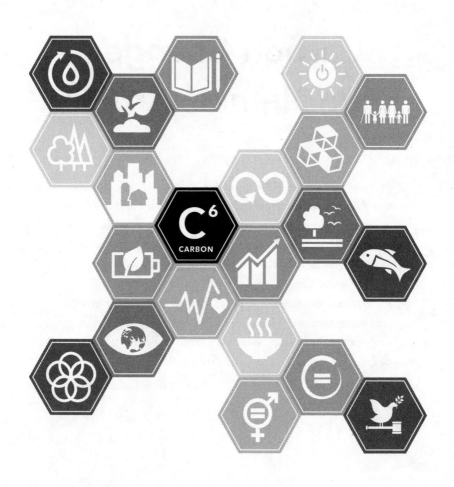

CHAPTER 1

Charles Keeling's Curve

At twenty-six, Charles David Keeling was fond of hiking the windswept cedar coast at Big Sur. He was doing postdoctoral work in geochemistry at Caltech. The year was 1954.

Keeling was someone who enjoyed rules, something that made him a good natural scientist. His son Ralph recalls that when mowing the lawn, Keeling insisted on following the habits of the previous owner of their home north of San Diego, an Englishman who had taken pride in his garden. And so Keeling maintained a precise two-inch strip between the sidewalk and the grass.

"It took a lot of work to maintain this attractive gap," said Ralph Keeling, who at thirteen had to do the mowing. But his father believed "that was just the right way to do it, and if you didn't do that, you were cutting corners. It was a moral breach."

On a hilltop at Big Sur in 1954, the twenty-six-year-old senior Keeling set up an apparatus to measure CO_2 in the atmosphere. Many at that time wondered whether CO_2 in the air was rising due to fossil fuels, but nobody had been able to take accurate measurements. Keeling developed techniques of great precision. His first readings at Big Sur gave the number 310. For every million parts of air, 310 parts were CO_2.

Keeling's work attracted the attention of Roger Revelle, director of the Scripps Institution of Oceanography. At that time, Revelle was working on a paper with Hans Suess suggesting that the Earth's oceans absorb excess anthropogenic CO_2 at a much slower rate than geoscientists had previously predicted, and that the shortfall would result in global warming. But Revelle needed better data, so in 1956 he recruited Keeling out of Caltech to work for Scripps and later sent him to Hawaii to set up the Mauna Loa Observatory.

The north slope of an active volcano might seem like an odd place for a research laboratory, but Hawaii's Mauna Loa had a bunch of things going for it: For starters, the military had already built a rough road to the summit. Plus, "the undisturbed air, remote location, and minimal influences of vegetation and human activity . . . are ideal for monitoring constituents in the atmosphere that can cause climate change."[1] While contamination from the volcano and atmospheric nuclear tests was sometimes detected, it could be identified as such and accounted for.

Keeling's first discovery at Mauna Loa was that CO_2 levels oscillated slightly depending on the season, with lower levels in the summer and higher levels in the winter. The reason? Most of the world's land is in the northern hemisphere, and plants there take in CO_2 as they sprout leaves and grow during the northern summer and then shed CO_2 as the leaves die and decay in the winter. Keeling was essentially measuring the cadence of Gaia's breath.

After a few years, Keeling could see another trend. Each year, the peak level was a little higher than the year before. Carbon dioxide was indeed rising, and quickly. Later tests by Keeling and others showed the increase was due not to volcanoes or other natural phenomena, but to the combustion of fossil fuels. The graph representing this came to be known as the Keeling Curve and is today etched into a marble wall at the National Academy of Sciences, fronting the Washington Monument.

Keeling reached out to Roger Revelle. Sobered by his own work on climate change, Revelle had left Scripps in the early 1960s and founded the now defunct Center for Population Studies at Harvard University, signaling his belief about the underlying driver of rising levels of atmospheric CO_2. Revelle tried to get President Kennedy's attention with Keeling's graph, to no avail. Following Kennedy's assassination, Revelle authored a report with Wallace Broecker, Charles Keeling, Harmon Craig, and Joseph Smagorinsky that managed to reach Lyndon Johnson. In a televised address, President Johnson told the nation, "This generation has altered the composition of the atmosphere on a global scale through . . . a steady increase in carbon dioxide from the burning of fossil fuels."[2]

After Johnson left office, the science reports landed on the desk of Richard Nixon, who couldn't have cared less. Then they landed on the desk of Gerald Ford, whose chief of staff, Dick Cheney, was a coal man. After that, they landed with Jimmy Carter, who leaned in. He read them. He asked questions. He, too, spoke to the nation, trying to explain the inexorable consequences of the exponential function and devotion to growth as antithetical to survival. He commissioned more scientific studies. Carter's science reports then fell upon the desk of a new president, Ronald Reagan, who disproved the notion that no one could care less than Nixon. Reagan busied himself removing the solar collectors Jimmy Carter had installed on the roof of the White House.

When George H. W. Bush, an oil magnate, became president, he was unnaturally interested.[3] But number 41 was interested for a different reason than Johnson or Carter. Bush knew that if Revelle and Keeling were believed, it could sound the death knell for the business upon which his family fortune rested. Bush reached out to the right-wing George C. Marshall Institute, which had been created to argue for President Reagan's Strategic Defense Initiative, and asked them to debunk the science of climate change.

The conservative think tank dutifully reported that the science behind global warming was inconclusive and "certainly didn't warrant imposing mandatory limits on greenhouse-gas emissions."[4] They had successfully used the same argument to muddy the waters in the tobacco controversy some years earlier, and were handsomely rewarded for it. The institute crafted the denialist position on human-induced climate change and pushed its agenda into if not a complicit at least a compliant media echo chamber.

Meanwhile, during his wilderness years at Harvard, Revelle was teaching physics to a US senator's son from Tennessee: the young Al Gore. Revelle's concern for the climate, it turned out, would have a far more profound effect on this curious theology student than it had on the series of presidents he so desperately tried to reach with his message. When Bill Clinton took office in 1993, he famously said, "You can never get elected by promising people less." He embraced fossil energy as part of his "all of the above" approach. Vice President Gore's plan to place a tax on carbon was defeated by Congress after the Global Climate Coalition, an astroturf front for the American Petroleum Institute, invested $1.8 million in a disinformation campaign. The landmark international treaty Gore negotiated in Kyoto was never forwarded to the Senate for ratification. Revelle was thwarted once more.

The Scripps program is now run by Ralph Keeling, who became a renowned atmospheric scientist in his own right and took over the measurement program from his father. In 2010, he predicted that the atmosphere would blow through the 400 ppm mark and not even slow down. He was right. We are now at 410. It took 250 years to burn the first half trillion tons

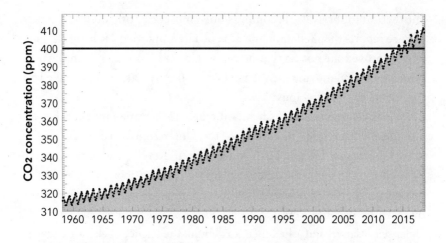

This graph shows carbon dioxide concentration as measured at Mauna Loa Observatory beginning in 1958 with Charles Keeling's early data collection extending through present day, based on data collection done by, among others, his son, Ralph Keeling, who eventually took over his father's post at Scripps Institution of Oceanography. *C. D. Keeling et al., "Exchanges of Atmospheric CO_2 and $^{13}CO_2$ with the Terrestrial Biosphere and Oceans from 1978 to 2000." I. Global Aspects, SIO Reference Series, No. 01–06, Scripps Institution of Oceanography, San Diego, 2001.*

of fossil fuels. At the pace we are going, our collective foot still mashing the accelerator, we'll burn the next half trillion by mid-century.

"When I go see things with my children, I let them know they might not be around when they're older," Ralph Keeling told the *New York Times* in 2010. "'Go enjoy these beautiful forests before they disappear. Go enjoy the glaciers in these parks because they won't be around.' It's basically taking note of what we have, and appreciating it, and saying goodbye to it."

A few years ago, the Oxford-based author Mark Lynas published his sensational book *Six Degrees: Our Future on a Hotter Planet*, in which he mapped out the consequences of each degree increase predicted by a 2001 Intergovernmental Panel on Climate Change (IPCC) report stating that average global surface temperatures could be expected to rise between 1.4°C and 5.8°C (3.5–10.4°F) by 2100. The book was largely an attempt to examine what would happen at the upper end of that prediction and resulted in the first Four Degrees and Beyond International Climate Conference, which took place in September 2009 at Oxford, in the United

Kingdom, and included about 140 delegates from science, government, nongovernmental organizations, and the private sector. Their conclusion? Four degrees might be survivable, with luck, by small, resilient, and ingenious groups of humans, like the Inuit or the Kuna, but it would never be survivable for civilization or a human population of any scale.[5]

There is presently a 93 percent chance that the planet will be more than 4°C (8.5°F) warmer than it is now by 2100.[6]

––––––––––––

To most people, four degrees may not seem like much. We typically experience that type of variation in an average day, immediately after sunrise or sunset. However, our skin is not the skin of the Earth. A four-degree global change involves heating or cooling vast volumes of land, air, and ocean. Four degrees is the difference between Earth's most recent ice age and where we are today.

There are seminars and conferences happening almost continuously now somewhere in the world as governments, businesses, the poor, the wealthy, and the academically minded try to come to grips with the existential threat posed by climate change. In November 2017, we attended the twenty-third UN Conference of Parties to the Framework Convention on Climate Change (COP-23) in Bonn, Germany. This UN event, besides crunching through a formal agenda to develop and enhance international agreements, provides an important time each year for sharing among the science community, policymakers and government officials, intergovernmental agencies, solution-oriented providers in business and finance, indigenous cultures, faith-based groups, environmental groups, and human rights organizations.

At a COP-23 panel discussion, Christiana Figueres, former head of the UN Framework Convention on Climate Change (UNFCCC) and principal midwife of the Paris Agreement, opened with the good news: $130 billion in green bonds were issued this year and there's been a dramatic price drop in clean energy. At less than 5 cents per kilowatt, its CAPEX (capital cost) is now below OPEX (operating cost) for even the newest fossil plants. In other words, there is really no economic excuse for not halting pipelines and coal trains and no business case to be made for either new or existing nuclear, oil, coal, or gas.

That is a sea change. Hans Joachim Schellnhuber, director of the Potsdam Institute for Climate Impact Research, quipped: "The world is awash with money. You park it in paradise, or Panama, whatever, you know? People have so much money they do not even know what to do with it, except avoiding taxes of course. That's a big sport, huh? But otherwise what to do with it? So private money is available if one were to set up a public/private transformation fund."

It was left to Kevin Anderson, deputy director of the UK-based Tyndall Centre for Climate Change Research, to deliver the bad news. He reminded the audience that emissions had been steadily rising since 1990. That's twenty-seven years of abject failure for multilateral UN negotiations. In 2016, emissions rose by a breathtaking 2 percent.

Anderson, a lean British engineering professor with a decade of industrial experience, principally in the petrochemical industry, has been struggling to engineer a way through the impasses of UN climate meetings for more than ten years. Loath to fly, he regularly takes trains and ferries between his home in Manchester and the Stockholm Resilience Centre in Sweden, where he is a visiting climate researcher. Anderson reminded the audience of something Pope Francis said in his *Laudato Si* encyclical: "The alliance of technology and economics ends up sidelining anything unrelated to its immediate interests . . . whereas any genuine attempt to introduce change is viewed as a nuisance based on romantic illusions."

"Where are those romantic illusions?" Anderson asked, "What are the real romantic illusions?" He listed them: a belief in naive and ephemeral textbook economics; an unshakable confidence in technical utopia ("As an engineer," he said, "I am quite prone to that one"); the deliberate neglect of time; faith in Machiavellian mathematics; and an implicit assumption that nature follows our rules.

"You'd think that after thirteen billion years," he remarked, "we'd know that nature will always win out."

Anderson then showed why many people in the climate policy world consider him to be rash. He challenged his own base. "Universities and NGOs have been corrupted by near-term power — we want to be at meetings in Davos, we want to be with the great and the good in our society, we fear questioning the dominant social paradigm — much more important than physics apparently — and we have a naive focus on particular pet technologies, whether it is nuclear, wind, or solar — it's always a supply technology."

In that canvas-walled temporary building in a city park in Bonn on that chilly November day, Anderson underscored his main point, the one we dare not ignore: We are not on track to achieve the principal aim of the Paris climate agreement, keeping global temperature rise to well below 2°C (4.2°F) while pursuing 1.5°C (3.2°F). Still greater ambition is required. Every few years, the IPCC produces a consensus report of thousands of scientists pronouncing what is technically required to stay within the safe operating boundaries of our climate system. Each report concludes that we are already in great danger and must act quickly.

As we know, we must bend the curve away from adding carbon to the atmosphere each year — currently a concave curve pointing up — toward subtracting carbon — a convex curve peaking out and trending down. Anderson said that to have a realistic chance of averting disaster, we need to reach an 11 percent decline rate per annum from 2036, to prevent catastrophic climate change above 2°C (4.2°F), or, better, a 20 percent decline rate from 2037, limiting ourselves to merely dangerous climate change at around 1.5°C (3.2°F).

This type of curve will not be easily achieved. An 11 percent decline rate is the inverse of doubling your fossil economy every seven years — so, halving it every seven years. Try to imagine half the number of commercial passenger flights in 2025 as today, or half the number of gas-powered engines. Half the number of Walmart Superstores bringing full cargo ships from Shenzhen to Houston. Then halve that by 2032 and again by 2039. You get the picture. Phasing out the worst fossil fuels in favor of the less evil renewable fuels will not bring carbon back into the safety zone fast enough. Anderson told the conferees that the IPCC got weak knees just thinking about that, so its Working Group III bent the curve back up a bit. The revised recommendation offers more attractive narratives, which are really just modest tweaks to business as usual.

How did IPCC justify staying above the required curve so long? Anderson said it did that by conjuring up nonexistent "negative emissions technologies" that will be sprinkled around the planet like fairy dust to reclaim, over the long term, the fossil emissions we are allowing ourselves in the short term (and indeed, subsidizing delivery of, to the tune of $5–$6 trillion per year,[7] or about $140 per ton of CO_2 emitted[8]).

Anderson told the audience that the amount of drawdown required for the fairy dust technologies is roughly equal to the natural carbon absorption capacity of Earth's biosphere. In other words, by 2040 or thereabouts we will need another Earth to absorb the extra pollution we'll be adding during the next twenty-two years. "The technology — we do not know if it will work, we are not really sure what it will look like, but we are relying on it already, in our own governments' position in this policy debate. . . . If these things don't work, we are locking ourselves into a three- to five-degree future."

Instead of that, Anderson proposed industrial countries reduce emissions at 10 percent per year or more, starting now, with the aim of a greater than 60 percent reduction in CO_2 emissions by 2025 and to be fully decarbonized in the energy sector by around 2035 to 2040. For the non-OECD countries, he extended the timeline for an extra decade, but fully decarbonized by 2050.

Hans Joachim Schellnhuber agreed we have a very short time to respond. If we waste time, we will have to bend the curve much more dramatically. We need to peak greenhouse gas emissions by 2020 and halve emissions every decade thereafter.

But nothing works in isolation, Johan Rockström, director of the Stockholm Resilience Centre, reminded the audience. "There is no such thing as delivering on Paris without the SDGs [Sustainable Development Goals] and there is no such thing as achieving the SDGs without succeeding on Paris — and there is very strong scientific support for this."

Rockström observed that we have entered a curve in the solar energy build-out where capacity doubles every 5.4 years. When you have only small numbers — all renewables are 2 percent or less of world energy capacity — that doubling doesn't seem like much, but follow that curve out a few more doublings. On the present trajectory the energy sector will achieve 100 percent renewables by 2045. We need the same kind of revolution in food production and buildings, but that will be coming.

"And if we do all of this — an energy revolution, an agricultural revolution, a technology revolution and a sustainability revolution — we have a 66 percent chance of staying under 2 degrees," Rockström said, but this was immediately questioned by Kevin Anderson.

Anderson: "We are really talking about shifting the productive capacity of society from what it does today and for at least the next 50 years to responding to the climate challenge. Renewables are really important, but renewables have really only been in addition to other fuels so far, they have

The UN Sustainable Development Goals

The eight Millennium Development Goals (MDGs), adopted at the World Summit on Sustainable Development in 2000 by 189 nations, despite the absence of any legally binding framework or finance mechanism, had notable successes, such as halving the number of people living in abject poverty (defined as living on less than $1.25 a day). When these goals expired in 2015, the UN meeting at Rio+20 produced a successor framework, the Sustainable Development Goals (SDGs), with a 2030 target date. The overarching aims of the seventeen SDGs are poverty elimination, sustainable lifestyles for all, and a stable, resilient, planetary life-support system.

Defining human sustainability in terms of ecological stability (sea level, stratospheric ozone, air pollution, forests, wetlands, coral reefs, eutrophication, temperature, ice sheets, carbon sinks, etc.) was both a scientific and social turning point. From the SDGs flowed environmental priorities, including nine boundaries that should not be transgressed without risking long-term human prosperity, or even survival.

Local conditions and aspirations play large roles in determining how individual governments or agencies respond to these targets. From the experience of the MDGs, it is important to have clear goals and measurability. Many of the SDGs, when tackled in an integrated way, deliver the desired outcomes for each as part of a greater whole. With more unified environmental and socioeconomic framing, reform efforts can also find the leverage to overcome obstacles and produce better and more durable results.

Many of the SDG targets have already been embedded in international agreements, such as the Paris Agreement on climate change.

not been substituting, and the climate does not care about renewables — of course, I am very much in favor of them. It does not care about energy efficiency. All it cares about is how much carbon you are putting into it.

And therefore we have to close down the incumbents. At the moment, every single year we burn more peat, more oil, more gas, more coal. We have never in human history, post–Industrial Revolution, seen a substitution in energy types at the global level."

While global anthropogenic emissions may be expected to decline now that almost all countries have ratified the Paris Agreement, they have not declined yet, and it takes a while for the results to appear. To still accomplish a rapid phasedown of fossil fuel emissions, the targets agreed to in Paris now *require* "negative emissions," meaning extraction of CO_2 from the atmosphere, or drawdown. To have any chance of averting catastrophe, drawdown must be ramped up quickly — to 4 billion tons of CO_2 annually by 2040 and 10 billion tons by 2050.[9]

Full implementation of the Paris Agreement will require a shift of global industrial economies no less profound than that which transformed Western and Soviet manufacturing in 1940. For the owners of factories, steel mills, mines, and myriad enterprises, a paradigm shift would have to transform their holdings from carbon polluting to carbon trapping almost overnight. In 1940, confronted with Nazism sweeping across Europe, entire industrial economies switched instantly from producing consumer goods to providing war materials. This transformation was not viewed as a trade-off. It was viewed as a response to a visceral threat to the future of great swaths of humanity. We face an even greater threat now. Can we repeat this rapid transformation to fend off the more worrisome effects of climate change?

What is needed now, today, is a low-cost, rapidly deployable, hugely scalable approach to carbon sequestration that could find political and social acceptance quickly, without the requirement of carbon taxes or offset markets. Indeed, it should scale quickly, use tested, off-the-shelf technology, be antifragile, employ low-skilled workers, and not endanger ecosystems or impoverish individuals, sectors, or countries. Nearly all proposed technologies are too far out, too outlandish, too unproven, maybe even too dangerous when we consider what unintended consequences might come along with them. Every technology with the exception of biochar, that is.

CHAPTER 2

X-Axis Over Y

If we wait just another 8 years, until 2025, then we are basically facing a cliff and we won't make the Paris goals anymore. The only way out of this is a massive amount of negative emissions — removal of CO_2 from the atmosphere. I'll tell you, I will believe it when I see it, that this is going to happen. I don't think that we will be able to do that at any scale.

— STEFAN RAHMSTORF, 2017

In mathematics the inflection point is the point on a curve where the sign of the curve changes. In business, the strategic inflection point is when massive change occurs or, as Andy Grove, Intel's cofounder, said, "an event that changes the way we think and act." It could be positive and lead to some dreamed-of rocket ride to success or it could just as easily be negative and lead to a company's or project's demise. To rebalance the carbon cycle, we need to find the inflection point on the Keeling Curve. Looking at its sawtooth pattern, we want to shorten the back edge and lengthen the lead edge of each tooth.

The two-degree Paris Agreement target requires "a balance between anthropogenic emissions by sources and removals by sinks of greenhouse gases in the second half of the century." This can be deceptive. To achieve a balance consistent with the two-degree target, merely dropping to zero fossil fuel emissions would not be enough, even if it were practicable (and very likely it is not). A balance where positive and negative emissions only canceled each other out also would not remove enough of the legacy emissions to hold to the two-degree line in the second half of the century. Staying below two degrees can only be achieved by carbon dioxide removal (or drawdown).[1] There are five generally recognized categories of these "negative emissions technologies."

1. Changes to land use management
2. Accelerated weathering
3. Marine biota
4. Direct air capture (DAC)
5. Biomass energy with carbon capture and storage (BECCS)

Lumped within BECCS, you'll sometimes also see "pyrolysis carbon capture and storage," or PyCCS, and this is the edge of the territory where things start to get really interesting to us. While BECCS (often rightly) is associated with clearing or otherwise sacrificing land to grow biomass crops and dangerous schemes to store carbon in, say, underground reservoirs vulnerable to earthquakes or leakage, the (low-tech, low-cost) transformation of underutilized organic carbon into a solid form that no longer is able to escape into the atmosphere should be at the center of any climate discussion.

First, a Word about Geoengineering

Attempting to manage the global commons to accommodate the outsized imprint of humans has led to a field of science called geoengineering. Many of the techniques lumped under that label are actually nature's own way of recovering when systems get out of balance. Others are entirely human-made, with a potential for very dangerous and damaging unintended consequences.

Although the idea of engineering the environment to counter anthropogenic climate change can be traced back at least to the 1960s, there is as yet no agreed-upon definition of *geoengineering* among research, policy, and civil communities.[2] In 2009, the Royal Society of the United Kingdom defined geoengineering as "the deliberate large-scale intervention in the Earth's climate system, in order to moderate global warming" and divided it into two classes: (1) solar radiation management (SRM), which is the intentional modification of the Earth's shortwave radiative budget by, for example, injecting stratospheric aerosols or painting roofs white to reflect sunlight to space; and (2) carbon dioxide removal (CDR) involving technologies that aim to reduce atmospheric carbon by increasing carbon sinks. CDR methods include large-scale afforestation, ocean fertilization, and mechanized capture of carbon dioxide.[3]

That said, we hope it goes without saying that this is an area where an abundance of caution is warranted because many geoengineering schemes risk further climate destabilization, may come at incredible cost, and could place us on a dangerous treadmill that each year saps more energy resources just to run in place but where any interruption, for any reason, could have cataclysmic consequences. Many proposed geoengineering technologies seem to assume that we can continue business as usual with a linear and extractive economic model of endless growth. Others, such as albedo enhancement by deforestation, ocean fertilization, or cloud seeding, are low enough in cost and accessible enough with moderate technical skills that they risk being launched at global scale by "wildcat" geopirates or other non-state actors.

While there are legitimate natural solutions that have been tossed into the geoengineering milieu, singling them out and categorizing their safety and social costs is tricky.[4] One thing we can say is that the last thing anyone should be doing is risking further destabilization of the climate with exorbitant, fragile, experimental techno-fixes. This is especially true when we need look no farther than the methods used by nature herself, such as water, trees, and soil, to heal damage such as this.

Afforestation and Reforestation

Drawing down carbon through changes in land use management can come in a variety of forms, from afforestation and reforestation to deliberately increasing or preserving carbon stocks in soils.

It is hard to perfectly quantify something with so many variables, but removing 10 percent of the CO_2 we add to the atmosphere every year through afforestation and reforestation would require conversion of between 1.2 million and 3.7 million square miles (between 3.1 million and 9.6 million square kilometers) of land.[5] Conversion of field to forest requires energy, labor, water, and other inputs. Nutrient requirements would be substantial (0.1–1.0 million tons per year of nitrogen, and 0.22–0.99 million tons per year of phosphorus) because you can't get plants to store carbon unless they also have adequate access to water, nitrogen, phosphorus, and other nutrients.

The loss of soil nutrient stores from deforestation and degradation, not to mention climate change, may make it difficult to reestablish healthy and growing forests in many places. If nitrogen fertilizers are used in nurseries or in tree-planting, that requires energy and fossil feedstocks (ammonia from natural gas) and increases emissions of nitrous oxide with its high global warming potential, three hundred times greater than CO_2. Even without accounting for the energy cost of fertilizer, the negative impacts of nitrous oxide, another potent greenhouse forcing agent, would exceed the beneficial effects of CO_2 reduction by 310 percent.[6] Alternatively, compost and biochar, managed correctly, could more than offset labor, land, capital, and operating costs[7] and also significantly reduce nitrous oxide (N_2O) emissions.[8]

If forests were planted in high latitudes or deserts, particularly where there is snow cover, the benefits of carbon drawdown could be offset by the loss of albedo, the ability of land to reflect light back to space. Climate scientists have known for some time that there are some places where putting new forests is a bad idea if you want to reverse climate change.

Net carbon uptake from the atmosphere does not follow immediately after replanting trees, and disturbances due to planting may release soil carbon that can take several years or even decades for the growing trees to offset. Maximum uptake is often in the middle growth period (twenty to sixty years depending on species, location, and local conditions), after which the rate of absorption starts to slow, although the amount of captured carbon (carbon stocks) continues to rise.

Captured carbon may be vulnerable to changes in local cultural and political priorities that might result in felling or the use of forest biomass for energy, as well as illegal logging, which is a major contributor to current rates of deforestation. In a sudden finance or energy crunch, new woodlots and forests would likely be seen as readily harvested "stopgaps," ignoring the long-term consequences. Landowners would have substantial financial incentives to liquidate those "assets." Also, newly established stocks, especially in marginal lands, may become vulnerable to widespread forest disease or fires, factors expected to be exacerbated by climate change.

While it is unlikely that the Paris goals can be reached without large-scale afforestation and while we are extremely strong proponents of "ecoforestry" (see chapter 4), it is also clear from the calculable carbon yields and available land limitations that forests alone will not achieve the Paris goals.[9]

Land Management to Increase Carbon in Soils

Many cultivated soils have lost 50–70 percent of their original carbon.[10] It is possible to partly reverse that loss by better management, such as cover crops, leaving crop residues to decay in the field, applying manure and compost, using organic low- or no-till systems, and employing other techniques to stabilize soil structure such as silviculture, terracing, keyline, and contour cropping.[11] The mitigation potential of such approaches has been estimated to be 0.07–0.7 tons of CO_2-e per year per hectare (0.173–1.73 per acre), with the greatest potential in warm and moist climates.[12] Applied globally to lands now under cultivation, drawdown potential is estimated at 1.3 GtC/yr, less than 10 percent of current global greenhouse gas emissions.[13]

For some of these practices, there is no added labor, capital, or energy cost to this change of practices. Indeed, good land management often decreases costs while increasing revenue. Such efforts can be sustained for decades or centuries, and eventually soil carbon content will tip into a steady state. However, such measures are easily reversed if farming returns to more intensive methods, such as by famine-driven desperation or generational loss of knowledge of good management practices.

Swamps along the edges of rivers and lakes and coastal ecosystems such as salt marshes and mangroves have historically stored large amounts of carbon, which has been lost through their removal. Coastal ecosystems are also especially vulnerable to rapid climate change.[14] Habitat restoration requires expenditure of labor, capital, and energy, and although these costs are not without compensating benefits such as improved fisheries, flood and storm damage prevention, and biodiversity, they tend to be difficult to quantify.

Project Drawdown, a consortium of more than two hundred scientists assembled by entrepreneur, environmentalist, and author Paul Hawken, estimates that establishing timber plantations on an additional 204 million acres of marginal lands could sequester 18.1 gigatons of CO_2 by 2050. At a cost of $29 billion to implement, this additional area of timber plantations could produce a net profit for land managers of over $392 billion. Project Drawdown also calculates that if 57 million acres of presently endangered coastal wetlands were protected, the avoided emissions and continued sequestration could draw down 3.2 gigatons of CO_2 by 2050.[15]

Accelerated Weathering

The amount of suitable and readily available mineral silicates (peridotite, feldspar, serpentine, pyroxene, and olivine, for instance) far exceed requirements for sequestering all conceivable anthropogenic CO_2 emissions.[16] An organization in the Netherlands has already been established to promote the dispersal of olivine to shallow marine environments.[17] These techniques, however, require mining, grinding, transport, and dispersal, with energy and environmental impacts proportional to scale.[18] It remains to be seen if this form of geoengineering can be economically implemented.

Another technique is in situ carbonation: injecting CO_2 into silicate rock formations, where the gas reacts to form stable carbonates. At the end of 2017, a facility in Hellisheidi, Iceland, began operations using the "free" energy of a geothermal power plant to capture CO_2 from ambient air, flush the filters into water, and pump it more than 700 meters underground. There, the CO_2 reacts with the basaltic bedrock to form solid calcium carbonate (limestone) after about two years. At full operational scale, the system is expected to capture 50 metric tons of CO_2 annually. The cost is in excess of $10 million.

Fifty tons is about the output of a single US household in one year. The cost of each ton, even amortized over a century, will run into the thousands of dollars, and needless to say, there are only a limited number of places on Earth where such a scheme could even be implemented. Work is under way to develop new approaches, often referred to as "enhanced weathering," and to bring costs down. If costs fall enough to compete favorably with other negative emissions technologies, or to become products and services provided as part of larger cascades, we can expect to see many of these approaches find commercial niches in the not-too-distant future.[19]

Rock weathering is the major source of most plant nutrients. Depending on the degree to which soils have lost nutrients from leaching and crop export, it can be economically beneficial for farmers and herders to amend soils with rock powders that will slowly release natural materials for decades. These can provide all the nutrient elements plants need except for nitrogen, and make them directly available to plant root mycorrhizae. Mixed with biochar to retain nutrients and water, they become slow-release fertilizers. The requirement for a continuing supply of minerals to sustain soil fertility supplies a robust motive for smallholder farmers and ranchers to devote the energy to accomplish this strategy, even in the absence of a fossil economy.

Marine Biota

Ocean fertilization using nutrients such as calcium, nitrogen, iron, and phosphorus is feasible, but like accelerated weathering it's unlikely to be practicable absent a viable revenue model, owing to the high cost of the mineral fertilizer quantities required. That viable revenue model could be provided by what is called marine permaculture.[20]

Marine permaculture deploys arrays, or lightweight carbon-lattice structures roughly half a square mile in size, submerged 8 feet below sea level, to which kelp can attach. Attached buoys rise and fall with the waves, powering pumps that bring up colder, nutrient-rich waters from far below. Kelp soak up the nutrients and grow, establishing a trophic cascade rich in plant and animal life. The kelp leaves can be harvested for food, feed, fertilizer, fiber, and biofuels, which more than offsets the costs of establishment. As the kelp forests expand, they cool the waters, which encourages phytoplankton blooms at the surface, as does the natural fertilizer from the new biotic community that forms in these submarine forests.

Plants that are not consumed die off and drop into the deep sea, sequestering carbon for centuries in the form of dissolved carbon and carbonates. Floating kelp forests could sequester billions of tons of CO_2, beyond that provided by the plankton. They also give life back to the oceans. According to Climate Foundation's Brian Von Herzen, "in just 57 hours after deployment, the system sparked plankton growth. Shortly thereafter, these blooms attracted various species of fish. Two weeks later, a 17-foot long (5-meter) whale shark was circling the area feeding on plankton that had started blooming."[21]

One of the effects of warming oceans is a plague of seaweed fouling beaches in tropical climates throughout the world. Red tides of toxic algae are killing fish and threatening swimmers. Instead, these forests of seagrasses, algae, and plankton could be sequestering gigatons of greenhouse gases annually, at a profit, if cultivated, harvested, and processed for their many benefits. We will talk more about this in chapter 11.

Direct Air Capture (DAC)

Direct air capture offers the possibility to capture CO_2 for economic benefit, but it is unlikely this can be said of the first generation of DAC installations, which employ amide sorbents to bind CO_2 from ambient air and concentrate from less than 0.001 percent (410 parts per million by volume, or ppmv) to greater than 95 percent (950,000 ppmv). The amides are scrubbed to recover the substrate – a process requiring high temperatures – and the CO_2 is cooled and transported to storage, more processes demanding energy. To minimize transport costs DACs can be sited close to CO_2 storage sites, but they still require significant energy and/or heat to operate. Current costs of the amide technology range $200–$1,000 per long ton of CO_2 compared with $30–$100 per ton of CO_2 for capture from flue gas.[22] The difference relates primarily to the degree of CO_2 concentration at the source. There are some limited agricultural and industrial markets for pressurized, high-purity CO_2 – fireproofing, beverage carbonization, enhanced oil recovery, industrial gas manufacturing, and refrigeration, for instance – but many of these depend themselves upon an industrial economic infrastructure that will be compromised by the phaseout of fossil energy. Moreover, selling the harvested CO_2 instead of storing it is a game of catch-and-release to the atmosphere, rather than drawdown.

If you need CO_2 at 95–99 percent purity, it will cost you much more than capturing 1 percent CO_2. This fact inspired Klaus Lackner at Arizona State University's Center for Negative Carbon Emissions to begin experimenting with using captured CO_2 to feed algae reactors and produce biomass and hydrogen.[23] In harvesting low-concentration (less than 2 percent) CO_2 and feeding that to tanks of algae for production of food, fiber, medicines, biogas, and chemicals, algal DAC meets the requirements of ecosystemic-benign, low-cost, socially acceptable technology with adequate incentives to function in the absence of the petroleum economy.

Algal DAC is flexible enough to run on most types of renewable energy, including its own biogas. It could supply algae bagasse to a BECCS system. It could be adapted to extract CO_2 from seawater using membranes. Anaerobic digestion of cellulosic wastes and biosolids with microalgal absorption of the carbon could make the business case: a favorable return on investment in both energy and financial terms without the problem of leakage. However, unless converted to biochar, the withdrawal is still temporary because the CO_2 remains in labile form and returns to the carbon cycle either as food or fuel. Combining anaerobic digestion, algal DAC, BECCS, and biochar would provide a whole solution. We'll look more at this in chapter 18.

Some types of DAC are flexible enough to run on most types of renewable energy as well as fossil energy. Collectors could even be mounted on ocean platforms, such as abandoned oil rigs. Besides absorption and adsorption, it might be possible to extract CO_2 from seawater using membranes, allowing the ocean to then reabsorb CO_2 from the atmosphere.

In Gothenburg, Sweden, Chalmers University is experimenting with a four chemical-looping combustor concept. The pilot plant uses two combustors for gaseous and liquid fuels of 0.3 and 10 kilowatts (kW), and two combustors for solid fuels of 10 and 100 kW. At this writing (mid-2018) the four chemical-looping combustors have been in operation for 4,000 hours with gaseous, liquid, and solid fuels, using close to seventy different oxygen-carrier materials. The reactors are very similar to normal circulating fluidized beds commonly used for combustion of solid fuels, including biomass, so the development cost for the boiler system is expected to be small. As mentioned earlier, the problem with DAC is less about capture than storage. For amide-based systems, after binding the CO_2 from ambient air, the sorbent is regenerated, releasing high-purity CO_2 — but to where? There are some limited agricultural and industrial applications for pressurized CO_2, but many of these depend on the industrial economic infrastructure that may be compromised by the phaseout of fossil energy.

Experiments with coastal coal plants over the past twenty years raised concerns about plans to pump captured CO_2 into the ocean, because as pure CO_2, it would react with seawater to form carbonic acid, and the oceans are already dangerously acidifying. They are more acidic today than at any time in the past million years. Professor Phil Renforth at Cardiff University in Wales has been exploring the potential for capturing carbon while marginally increasing ocean alkalinity. He would do this by reacting CO_2 with minerals like forsterite, calcite, or portlandite to produce bicarbonate solutions that could be safely disposed of at sea, although not all in the same place.[24] While we are always hopeful that experiments such as these might make contributions to our growing solution portfolio, we want to be cautious. Humans have a fondness for breaking all the natural laws but one: the law of unintended consequences.

Until there is a commercially viable, antifragile way of sequestering carbon in gaseous or liquid form, DAC cannot be considered a safe and economical solution.

Biomass Energy/Pyrolysis Carbon Capture and Storage (BECCS or PyCCS)

The term *carbon capture and storage*, or CCS, began in the coal industry. It was the centerpiece of a massive public relations campaign, begun in the 1980s, to avoid the inevitable consequences to their industry of people awakening to the threat of climate change. The catchphrase was *clean coal*.

On August 22, 2017, President Donald Trump, speaking in Phoenix, told the gathered crowd, "We've ended the war on beautiful, clean coal. It's just been announced that a second, brand-new coal mine, where they're going to take out clean coal — meaning, they're taking out coal, [and] they're going to clean it — is opening in the state of Pennsylvania, the second one."

The president was referring to a new mine opened by Corsa Coal, but he got his facts confused. *Clean coal* refers to something that happens at the burning stage, not during mining, fuel processing, or transportation. Even at the end-of-pipe phase, however, it is fiction. At the same time the president was touting clean coal for Pennsylvania, a giant clean coal facility in Mississippi owned by the Southern Company was shutting down, years behind schedule and $4 billion over budget. A $4 billion Texas Clean Energy Project coal gasification plant had met the same fate in August 2016. And prior to that, funding was pulled from a comparable FutureGen project in Illinois in 2015 after twelve years of work to get that plant running. In all, more than thirty such billion-dollar boondoggles litter the landscape, all victims of their own hype.

Scrubbing CO_2 out of an emission stream isn't particularly difficult, chemically speaking. Amines can adsorb carbon dioxide and hydrogen sulfide, but the process takes a huge amount of energy and space. To pull the CO_2 back out of the amide, it must go through a hot solvent bath and then the solvent-and-CO_2 mixture must be heated to remove CO_2 for compression, transport, and storage. The energy required to do this equals about one-third the output of a typical coal plant, and the area required might be equivalent to the size of the entire power plant — acres. Then there is the problem of storage.

For well-selected, -designed, and -managed geological storage sites, risks are said to be comparable to those associated with oil, gas, and coal mines, but this conclusion is premature and suspect. Injection of a gas into geological storage is not the same as removing long-interred crude oil and coal. We have no experience with long-term storage, and the rates of leakage witnessed in the short term do little to instill confidence.

Against this background comes the proposal, from the policy recommendations of IPCC and other distinguished bodies, for biomass energy carbon capture and storage (BECCS). They are not referring to a solid form of carbon, like coal, but rather to storing carbon as a compressed gas or liquid — CO_2. And the suggested means is exactly what had been proposed for clean coal — removal from flue gas by amide scrubber, reconstituting the pure CO_2, then transport and storage.

In the United States alone, estimates for the proposed ramp-up of biomass energy with carbon capture and storage are as high as 1 $GtCO_2$/yr in 2050, and between 1 and 3 gigatons annually by the end of the century. These estimates are limited less by availability of biomass than by suitable geologic storage sites and transportation.[25] With geologic storage, injection rates that exceed the injectivity of a particular storage reservoir increase subsurface pressures and may create fractures in the cap rock, induce earthquakes, or activate faults, making the project more prone to leakage, dangerous to nearby residents, and costlier to monitor.[26]

Fortunately, there is another path to BECCS that makes a better business case. Three Cornell University scientists, Dominic Woolf, Johannes Lehmann, and David R. Lee, have called this variation BEBCS (biomass energy with biochar capture and storage).[27] The distinction is that while BECCS achieves a net removal of CO_2 from the atmosphere by combusting biomass and geologically storing the gas, BEBCS achieves the same net removal by storing fixed carbon (biochar) in soil. As a cobenefit, biochar improves the fertility of degraded or less fertile soils, improves soil-water management, and supplies other services. Pyrolysis transforms labile, organic carbon into biochar or pyrolysate — the distinction has to do with feedstock and end-product purity — that is no longer a gas that wants to go back to the atmosphere or a liquid that can seep into groundwater. Pyrolysis produces heat and organic chemicals that have commercial value. An algal stage can be added to the bioreactor to increase the output of food, energy, and biochar.[28] The process, with cascade co-products, more than pays for labor, capital, and energy and does not need any fossil inputs.[29]

The primary feedstock for BEBCS is agricultural residue from maize, wheat, barley, oats, rice, cotton, sugarcane, and sorghum; woody biomass and woody residuals from forest thinning and sawmill residue; bagasse from production of sugar or ethanol; and dedicated energy crops, such as biomass sorghum, energy cane, eucalyptus, miscanthus, pine, poplar, switchgrass, and willow. (Caution is warranted with dedicated energy crops, and we address this in chapter 4.) Properly conditioned, the resulting biochar can be added to soil as a substitute for fertilizer. When applied to soil, it does not decompose to feed plants but instead becomes a long-term habitat for worms, bacteria, and fungi, increasing crop yields and adding fertility back to the soil. Municipal solid waste, food waste, manures, and biosolids are not included in what we are calling biochar because of contaminants and high moisture content.[30] In that case we have chosen the acronym PyCCS (pyrolytic carbon capture and storage) to refer to pyrolyzed biomass which may be unsuitable for use in soils used for growing food.

While earlier studies, looking only at BECCS and BEBCS, suggest an annual drawdown potential of 1.8–2.4 $GtCO_2$/yr, we, like others, have discovered this to be a significant underestimation when the full potential for BEBCS is reached and PyCCS is added.

The drawdown capacity for pyrolysates is at least an order of magnitude higher, because when you expand beyond soil storage as the end repository you also remove restrictions of moisture, minimum carbon thresholds, and contaminants. In the case of PyCCS, the repository could be concrete in bridges or asphalt in highways and the feedstocks can include biosolids from sewage treatment plants and livestock manure, or contaminated carbon products presently going to landfills.

Pyrolysis plants do not require the enormous finance and energy-intense exploration, physical structures, and transportation that have characterized the fossil fuel era. They can start very small and scale to whatever degree is economically viable within a geographical region. While constrained by the quality of soils, weather, and terrain, they are nonetheless viable and profitable enterprises even in the absence of capital, fossil fuels, and skilled labor. Biorefineries based on BEBCS of PyCCS are versatile, scalable, and shovel-ready.

Contrary to what some claim, there is no shortage of biomass. Fields of food do not need to be replaced with biomass energy crops. Our linear economy and lifestyle is positively drowning in wasted organics — sewage sludge, livestock manure, invasive species, green waste (yard clippings), food waste — the list goes on. Many are either landfilled or burned in an effort to ship them away, although as we are learning, there really is no "away" anymore. This type of handling comes at a cost, not just to waste producers who have to arrange disposal, but to those on the receiving end — those who live near landfills or close to areas being burned and despoiled.

We can have our energy *and* our food at the same time. We can get rid of landfills and incinerators, waste lagoons, and ocean dumps all at once. To do this, we need to transform our old linear model into a carbon cascade economy, in which a growing portion of underutilized labile carbon cycle will be converted into recalcitrant carbon (see chapter 4). Planting enough forests while simultaneously curbing emissions will get us into net drawdown territory. Preventing carbon from returning to the atmosphere by making biochar and pyrolysates an integral part of that forest-industry-energy mix is a necessary second step after pulling out carbon to begin with. We can begin to sequester carbon in concrete highways and highrises. We can grow kelp and, after pressing it for leaf protein, char that and build coral-restoring coastal filter barriers. Carbon abuse and waste becomes carbon abundance and recycling. The change we make starts to stay changed. Circular carbon economies begin to cascade. Carbon rebalancing can begin in earnest.

But the fact is that we'll never get there unless we revisit our antagonistic relationship to carbon. Right now carbon is getting a bad rap. Carbon creates dirty energy. Carbon creates grit, grime, and gunk. Carbon should be global warming enemy number one. But in truth, carbon is something we should all love and cherish. Carbon is life. Out of balance, carbon suppresses life. In the right balance, carbon provides life.

CHAPTER 3

God's Own Atom

I have no doubt that in reality the future will be vastly more surprising than anything I can imagine. Now my own suspicion is that the Universe is not only queerer than we suppose, but queerer than we can suppose.

—*J. B. S. Haldane, 1927*

Although Joni Mitchell sang, "We are stardust, we are billion-year-old carbon," we are actually — not to burst anyone's bubble — the ashes of helium.

A single carbon atom was too large an object to form in the beginning, during the Big Bang. Its birth had to wait half a billion years, until subatomic particles slowed, cooled, coalesced, and formed the earliest stars. And then it had to wait billions more years, as those stars brought the swirling proton gases into stellar nucleosynthesis, a process that continues today.

Every time four hydrogen atoms fuse to form one helium atom, two positrons (antielectrons), two gamma rays, and two electron neutrinos (in one of three "flavors") burst across space, heralding the birth of carbon. Eventually stars will use up their hydrogen and begin to cool, change color, and expand into red giants. At their cores, helium is compressed until the forces are strong enough to begin fusing nuclei (proton–neutron pairs or "alpha particles") together to form larger atoms. This is the way we get carbon, oxygen, and iron.

Were it not for the death of stars, there could be no flesh and blood.

The improbability of this occurrence in the beating heart of dying stars creates a mystery for science. We cannot mathematically explain the abundance of carbon in the universe. According to one calculation, the energy level of this excited state must be extremely precise, within two frequencies of light — a condition called triple alpha — or carbon would not occur.[1]

That degree of fine-tuning is a phenomenon called Hoyle resonance. The astrophysicist Fred Hoyle, who first argued for stellar origins of carbon in 1954, concluded by 1982 that the fine-tuning of the triple-alpha condition was evidence of a "superintellect." In other words, carbon is a signed work. The scientific community rejected Hoyle's intelligent design argument as unparsimonious, preferring to leave the matter of triple-alpha resonance to uncanny coincidence. Whatever the case, carbon brings to Earth some very special qualities. It well deserves to be a signed work of art.

Carbon-12, the most common isotope, has six protons, six neutrons, and six electrons. Its mass number is 12, atomic number 6. Because of its stability across many forms and combinations, it has become the standard relative to which the atomic weights of all the other elements are measured.

Arriving on the solar wind some billions of years in the past, carbon stardust lingered for a time in Earth's atmosphere before hitching a ride on a raindrop and falling into the ancient oceans. There it bonded with hydrogen to form some of the earliest chain molecules we call "organic." As these molecules formed nucleoproteins and began to reproduce themselves, the oceans came alive with a carbon food web. Carbon became a common denominator of all known life that followed. Single-celled organisms evolved to capture the energy of sunlight — photons expelled from the core of our nearest star in reactions such as the birth of helium — and used that energy to weld carbon with water to release oxygen and form the long carbohydrate rings we call sugars.

In the Paleoarchean, 3.5 billion years ago, carbon stardust in the form of single-celled cyanobacteria became the first plants, and then the first animals. Earth's atmosphere then was very different from what we breathe today. It was a toxic brew of methane, ammonia, and other poisonous gases. Cyanobacteria, able to take in methane and exhale oxygen, transformed the Earth's anaerobic (oxygen-poor) atmosphere to today's aerobic (oxygen-rich) atmosphere, permitting the evolution of life on land. Our oxygen-rich atmosphere was generated by the exponential growth of cyanobacteria during the Archaean and Proterozoic eras.

Unfortunately for carbon-breathing bacteria, the arrival of an oxygen atmosphere made living tough. You could say that the blue-green algae caused the first mass extinction. Asked if she thought the climate change being described by Al Gore would be catastrophic, National Medal of Science laureate Lynn Margulis told author Charles C. Mann, "Sad, sure. But a *catastrophe* — no. Oxygen — now *that* was a catastrophe!"

The algae (actually bacteria) adopted a clever strategy to survive. They amassed in thick colonies and enclosed themselves in a membrane, becoming stromatolites, the first multicell organisms. Sometime in the late Proterozoic, or in the early Cambrian, in an event known as endosymbiosis, they began to take up residence within certain eukaryote (algae) cells, producing food for their host in return for a secure home with minimal oxygen. The chloroplast with which plants make food for themselves is actually a blue-green algae living within the plant's cells. The photosynthetic process itself, characteristic of all green plants, is actually the work of spinning bacteria within the cell walls of the plant.

Today these algae, born of stardust, are still essential to life on Earth, and they are responsible for making the carbon cycle work. Where cold waters well up from the depths (such as in the North Atlantic), the rising water pulls carbon to the surface. That carbon is essential to form the bodies of plants (phytoplankton) that convert sunlight and carbon dioxide into food for animals (zooplankton). When these plants and animals die, their remains descend to the ocean depths, drawing carbon back down.

Carbon is stable and tetravalent — making four electrons available to form covalent (shared-electron) chemical bonds. The atoms of carbon can bond together in different ways, termed allotropes of carbon. Carbon can even form covalent bonds with other carbon atoms, which in turn can share electrons with others and so on, forming long strings, complex branchings, and "head-to-tail" rings of carbon atoms. There is practically no limit to the complexity of carbon branches or rings. Allotropes include diamonds, graphite, graphene, buckyballs, and carbon nanotubes.

Because of these features, carbon can bond with many elements — more elements than any other element. Soil and carbon consultant David Yarrow provides the example of vitamin B_{12} — the so-called vegetarian vitamin.

"B_{12}'s mineral cofactor cobalt has six valence electrons in outer orbitals and thus makes six bonds to other atoms," Yarrow tells his students. "B_{12} bacteria build a complex structure of carbon rings to enclose an atom of cobalt. Unlike chlorophyll and heme, whose carbon rings are in a 2D, flat

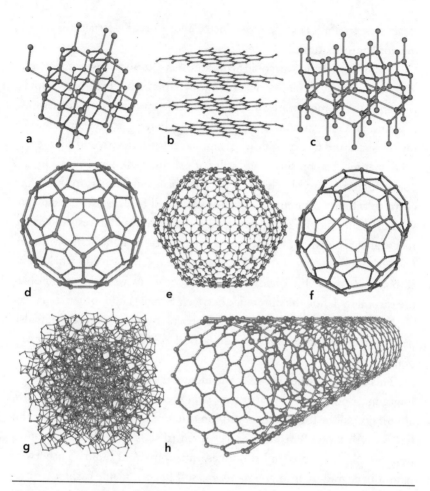

The most interesting characteristic of carbon is its nearly unmatched versatility. The eight allotropes (different molecular configurations) of carbon are (a) diamond, (b) graphite, (c) lonsdaleite, (d) C_{60} (Buckminsterfullerene or buckyball), (e) C_{540}, (f) C_{70}, (g) amorphous carbon, and (h) single-walled carbon nanotube or buckytube. *Design created by Michael Ströck.*

disk, B_{12}'s more complex carbon rings form a 3D structure. Thus, cobalt's magnetic energy is focused and harnessed to perform certain critical, universal energy exchange functions needed for DNA replication, hormone synthesis, and red blood cell formation."[2]

Because of its strength and versatility, carbon forms the backbone of biomolecules. Its tetravalence makes it a universal connector for complex structures. Its ability as a semiconductor allows organic structures to hold memory, information, and intelligence.

Without cell walls, we would all be aqueous solutions. Instead, our cells' water is enclosed in semipermeable thin chains of hydrocarbons (lipids) that create a cell membrane, through which are transferred carbohydrates, proteins, and information. Inside the cell, carbon is a traffic cop, organizing the space and directing flows of amino acids, electrons, and charge.

Sometimes carbon surrounds cells with shells, mesh fences, or skin and bones. It creates favorable habitats for microbial communities to thrive and coevolve within a community organism. Multicell families expand and conspire to create colonies of independent, interdependent cultures. And at the root of all that life is one essential, versatile element — carbon, God's own atom.

Carbon is everywhere. We are carbon beings on a carbon world. The carbon cycle is our life cycle. Dust to dust, ashes to ashes. We sometimes use carbon to define the difference between organic and inorganic materials. While carbon exists in all organic materials to varying degrees, it is also present in many inorganic compounds. Both forms have potential for carbon cascades.

This book describes many of the identities assumed by carbon but we linger most often upon what is now being referred to as biochar, a hard, almost crystalline form that persists long after most photosynthetic carbon has decayed away. Biochar may be added to soils with the intention to improve soil functions and to reduce emissions from biomass otherwise naturally degrading into greenhouse gases. Its formidable carbon sequestration properties can be measured and verified in a characterization scheme, or in a carbon emission offset protocol.

The term *biochar* appears more than seven hundred times in this book, although it only came into use at the end of the twentieth century after the discovery of charcoal-enhanced soils created by indigenous cultures dating back thousands of years. Scanning the research literature on biochar, now growing exponentially, it is clear the term has gone well beyond its original intent. It's conventional wisdom in the biochar community that not all biochar is the same and not all char is biochar. Rather than attempting to standardize it, "fit for purpose" is the latest way of classifying different types of biochar.

We also know that people have been making and using charcoal since at least the time of cave wall paintings sixty-five thousand years ago. As recently as the early nineteenth century, professional charcoal makers, known as colliers, were common. Their job was dirty and dangerous, but necessary because charcoal was essential for heating, cooking, blacksmithing, and metallurgy. The best charcoal was made at low temperatures. Fuel wood — often oak, hickory, ash, or maple — was generally stacked in piles and covered with damp earth, lit from the top of the pile, and left to combust and smolder for days. Burning wood slowly and at low temperatures is still one of the least expensive and easiest ways to make charcoal.

Biochar can be made from a much broader range of materials than charcoal can. Crop residues, manures, and wood are all potential feedstocks. In addition to use in the soil, newer uses for biochar are now competing with traditional uses for activated carbon, carbon black, and graphite. It is also well poised to displace, at least in part, noncarbon materials such as sand, Styrofoam, and fiberglass.

Tweaking the production parameters and the feedstock can result in widely divergent characteristics that allow biochars to be optimized or designed for specific uses. Perhaps the most defining characteristic of biochar versus many of its other carbon cousins (see appendix A) is its planetary impact. Unlike charcoal made from dwindling forests or activated carbon made from fossil fuels, biochar can have a tremendously positive effect on the planet both in reversing climate change and in providing products and services the world needs.

The specific thermal modification that converts biomass into biochar can be viewed from two closely related processes: pyrolysis and carbonization (see appendix B). Pyrolysis relates to the chemical breakdowns that result in the liberation of pyrolytic gases. Carbonization is what comes next: the chemical buildup of the carbon atoms into solid structures. The bulk of pyrolysis and carbonization reactions occur in the temperature range from about 320°C to 800°C (608°F to 1472°F). One can think of pyrolysis and carbonization as simultaneous physical–chemical processes, changing the biomass into pyrolytic gases and charcoal.[3]

We have occasionally found sources for perfectly good biochar already pyrolyzed and ready to micronize, moisten, mineralize, and microbialize. One biochar researcher we know uses her old aquarium filter charcoal to grow amazing tropical plants in pots. In rural Tennessee we have managed to arrange delivery of truckloads of filtration char from distillers like Jack

What Is Carbonization?

Not all forms of combustion result in carbonized by-products. Complete combustion produces ash, a low- to no-carbon, highly mineral substance. Incomplete combustion, on the other hand, can produce charcoal, which is about 95 percent carbon.

Think of the common matchstick. At one end there is a phosphorous igniter, which if struck briskly responds to the heat of friction and flashes into flame. As the phosphorus burns away, it ignites the higher-ignition-temperature wood or cardboard. What has really happened is that the heat of the burning phosphorus warmed the adjacent fuel until the elements with the lowest vaporization points began to volatilize — convert to gas — and then ignited in the presence of flame. The last element left behind — the one with the highest vaporization point in the fuel — is carbon.

As the match burns down, it burns away everything except carbon — leaving a black skeleton — until finally, it ignites the carbon (the heat serving as a catalyst to join carbon with oxygen). All that remains is ash — the mineral impurities of the fuel that never attained the high temperatures they require to vaporize. Suppose that we stop the match — blow it out — before it has a chance to turn to ash. What remains is that black carbon skeleton. Charcoal.

What we have done is "gasify" the fuel without allowing the carbon to join with oxygen and burn, which is what gasifying stoves and furnaces do — they allow the combustion of the low-temperature volatiles but restrict combustion of the high-temperature volatiles, primarily carbon. It doesn't matter whether the fuel is wood, plant residues, chicken litter (the sweepings from the coop), or biosolids. The process is the same, and the by-product is called biochar (as long as certain criteria are met in terms of carbon content, etc.)

Gasifiers can be used for other common carbon-rich materials, including shredded tires, pulverized plastics, or municipal landfill wastes, but we are cautious not to call that biochar. Although these feedstocks can be pyrolyzed, they might not exhibit soil fertility

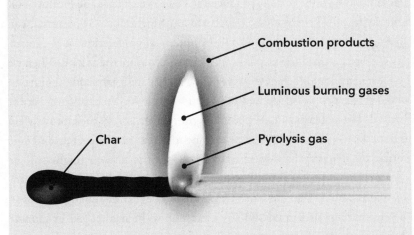

Combustion products

Luminous burning gases

Char

Pyrolysis gas

When a match burns, its biomass passes through stages of pyrolysis, carbonization, and oxidation.

properties and often they are detrimental. In this book we call them pyrolysates.

Although there is some disagreement in how carbonization and pyrolysis relate to combustion from a terminology perspective, they work together. Pyrolysis occurs in an oxygen-limited environment using very high temperatures, produces syngas and char (typically 25 percent by volume), and can be tapped for heat or electricity. Carbonization completes the process by modifying the chemical bonds within the biomass — dehydration, conversion from aliphatic into aromatic bonds, and consolidation into local graphene complexes — making them less accessible to living systems.

This diversity of chemical structures and overall greater bond stability thwarts the ability of living systems to supply appropriate enzymes to transform the carbonized bond structures. In a sense, carbonization converts biomass into a new, more stable form of carbon, which is more difficult to digest for the microbes — especially if there are sources of more palatable noncarbonized biomass available.

Daniels and George Dickel (produced in the town of Cascade Hollow). Of course, the whiskey residues, which are antifungal and antibacterial, can pose a problem for the compost pile or microbial inoculation, so we generally season this product outdoors and let rain denature the spirits. When we first learned about biochar, we were convinced its ability to sequester carbon would make people sit up and take notice. That hasn't happened yet for many different reasons. Later, we thought that its ability to improve yield and soil fertility was really what was going to cause people to sit up and take notice. That hasn't happened either. Both of those ideas might still be approaching some kind of inflection point, but when or how the great turning might happen remains a mystery. Maybe if regulatory changes occur in carbon markets or if food security continues to be challenged by climate change, that will force a paradigm shift to regenerative agriculture. Or maybe there is something else that could drive such a change.

Recently we've begun to see there might actually be other triggers related less to the intrinsic qualities of biochar than to the other harmful, wasteful, or expensive problems it can address. Here is one example. When biomass rots or burns, it creates air pollution. It gives off carbon either aerobically (combined with oxygen) as carbon monoxide (CO) or carbon dioxide (CO_2), or anaerobically (without oxygen) as methane (CH_4). Whatever other volatile compounds are in the biomass as it decays or burns likewise become gas and waft off into the atmosphere or are carried off as airborne particles of soot.

This type of air pollution can be a huge problem, as anyone who has tried to cook over a smoky fire or stood near an open sewer can attest. It is not just a problem of smoke or foul odors; it can do some serious damage to health. Almost eight million people die each year of chronic pneumonia from inhaled smoke from indoor cooking.

It is very common in many parts of the world to burn crop residues after harvest and processing, even when people know it is bad for their health and a nearly complete waste of soil nutrients. But what are farmers to do? Piling up agricultural or food-processing residues to rot or dumping them in gullies or landfills, besides contributing a dangerous greenhouse gas (methane), can attract rodents, produce foul odors, and cause water pollution.

If you take just those two examples — burning and rotting — they are precursor problems that could stimulate the next agricultural revolution.

Why burn waste biomass outdoors when you can just move it indoors and get paid? Pyrolysis sends part of the original plant material to the atmosphere, but much of the carbon is converted into a longer-lasting version of its former self. Now you have a product that can be added as a soil amendment to create structure; buffer against drought, fire, and flooding; and build the soil food web. Or, as we show in this book, it can be embedded in other long-lived products such as roads, bridges, and buildings. Using it to plant trees means survival, resilience, and growth of new forests. Using it to clean water means coastal mangrove wetlands can be preserved, coral reefs can be restored, and rivers can be restocked.

Because carbon takes a harder, crystal-like form during pyrolysis, it stays in the soil for thousands of years. The same organic waste turned into compost or mulch (or just left be) would be consumed by microbes and respired as CO_2 and N_2O in a few short years. Pyrolysis provides a unique advantage that helps turn our carbon foe into a long-lasting friend. Harvesting carbon to make maple syrup, banjos, carbon-fiber bicycles, and spaceships extends the life cycle of carbon. Returning carbon back to the earth, waiting to be reborn one thousand years or more from now, is a revolutionary act.

CHAPTER 4

Carbon Cascades

What you people call your natural resources, our people call our relatives.

— OREN LYONS, *Haudeneshaunee*

*L*ong *tail* is a term used to describe the residual earnings that come from an asset after it has had its initial market impact. In the case of a book like the one you are holding, the long tail might begin as a magazine series; move to ebooks, audiobooks, webinars, and films; and then pass along to a very long and fruitful second life in foreign translations and excerpted reprints.

Managed well, most forms of carbon can have very long tails. As labile carbon from photosynthesis, the tail moves from the soil food web and the exudates of fungi and bacteria into the roots, stems, and leaves of plants, or, in the case of marine life, from the nutrients brought to surface by whales to feed ocean gardens of algae and plankton. Some of this nutrient flow is consumed by animals and passed up a food chain until the organisms reach the end of their lives, die, and the carbon is returned to cycle around once again. There is no end destination, just a continuous cycle.

In a carbon cascade, a growing portion of the labile carbon cycle is diverted into recalcitrant carbon. In this more stable form it can pass through many useful stages as food, filter, fodder, or building materials before returning to the part of the cycle where the story began, the soil. During its transformation, useful services like heating, cooling, and power can also be generated. It can restore degraded lands and rebuild biodiversity. It can mitigate the effect of changing climate and ease or enable adaptation. Carbonized biomass might begin as soil nourishment for a small moringa tree, later becoming the twigs a family collects for daily cooking in a rural village (perhaps with stewed moringa leaves on the menu), before the family grinds the cinders for use in worm compost or once more as orchard fertilizer. In a cascade, carbon is retained, not squandered thoughtlessly. While it rebalances atmospheric and terrestrial carbon levels, it is also building a human culture better attuned to the rhythms and requirements of the natural world.

But wait, as they say, there's more. The more we rebalance the carbon cycle, the more we can lengthen its tail; in fact, the cascades amplify themselves, beyond what could even be described with a linear metaphor, and into a deeply circular one. Instead of the three-stone fire common in much of the world, perhaps the cook is a proud owner of a BioLite stove. The stove captures the heat during cooking and converts a portion to electricity to charge the family phone and/or provide home lighting. Maybe the cook doesn't need to go looking for twigs anymore because the same village cooperative that sold her the stove also planted fast-growing moringa trees and harvests not only the fuelwood but also the nutritious leaves, medicinal bark, and high-protein seedpods. And perhaps instead of adding the biochar to the compost pile, the family first uses it to filter their drinking water and then deodorize their bathroom before sprinkling it into the dry latrine where it removes odors and later makes an even better compost accelerator. Or perhaps the biochar is taken from the stove, ground into powder, and mixed with cattle feed, where it accelerates the growth of calves, raises milk production, and keeps the animals healthier. The animals deposit their biochar-containing manure in the pasture where it is taken underground by dung beetles, or in the barn where it is collected for fertilizer, and traded to a cooperative to speed the proliferation of the moringa trees.

In economic parlance, these cascading uses would be called multipliers. The more transactions, and the faster they occur, the more rapidly the standard of living of all the participants in the local economy grows. As the benefit of the captured carbon cascades throughout the system, there's also a cascading social impact. The two are, in fact, indivisible. Compare this with our current extractive linear model: The tail of coal, oil, and gas winds through exploration, test drilling, removal, refining, distribution, retail sales, burning for end use, and solid, liquid, and gaseous emissions, and ends in a contamination cascade of the soil, oceans, and atmosphere. There, but for the lethal legacy of climate change and latent cancers from inhaled or ingested toxins, the story of fossilized carbon typically ends. There is a branch tail that extends farther and lasts a bit longer. It winds from the refineries through petrochemicals to a wide variety of short-lived or long-lived products (e.g., plastics), employing a very large number of people along the way, before coming to rest, as did the main branch, in pollution of one type or another, possibly in the carcasses of billions of marine creatures.

Let's take another example: In Stockholm, Sweden, through years of trial, error, and persuasion, a small but dedicated team succeeded in building a strong, consistent, nonseasonal demand for biochar. Having proved the various benefits of using biochar in structured soils — one part biochar, one part compost, to six parts gravel by volume — the city found itself unable to meet its own demand. It had to import biochar from vendors in the United Kingdom and Germany by the boatload for the growing number of urban projects. That created an opportunity.

The Stockholm team first focused on improving urban tree survivability. As they expected, the trees and shrubs they planted in biochar-augmented soils thrived compared with trees in heavily compacted soils. Urban trees provide services estimated at more than $500 million for a city the size of Stockholm, including improved air quality and human health, production of oxygen, carbon capture and storage, wildlife habitat, reduced heat island effect, reduced energy demand, and improved property values. When Stockholm gets a heat wave, drought, plague of bark beetles, or deep winter freeze, its trees can now more easily survive. Not having to replace urban trees after each natural disaster represents an enormous cost savings to taxpayers.

Stockholm discovered an unintended but enormously valuable cobenefit. Switching to gravel-based, biochar-enhanced structured soils for urban landscaping provided a significant reduction in both the volume and contamination of stormwater going to the city's wastewater management system. That reduced municipal wastewater management costs. After winning the 2014 Bloomberg Mayor's Challenge — a competition for cities to come up with solutions to major urban challenges that can be adopted by other cities — Stockholm took the next step in creating a closed-loop, waste-to-biochar-plus-energy-production facility to eliminate the need to import biochar and upcycle urban green waste.

Simply to recycle is not enough, because that is usually a downward spiral through less valuable and less complex products to final disposal. When we speak of upcycling, we borrow that word from architect William McDonough and chemist Michael Braungart, authors of *Cradle to Cradle: Remaking the Way We Make Things.*[1] Upcycling is about the biological design of industrial processes, endowing each with its own metabolism, and collectively giving all processes and products the ability to function as ecosystems. Regenerative system theorist Daniel Christian Wahl says that being able to do this success-fully is the first, important step toward creating regenerative cultures.[2]

Opportunities for carbon cascades abound. Enormous amounts of water are used to process food. Grocers need it to remove dirt, leaves, skin, and seeds from fresh produce. Processors need water for blanching, cooking, cooling, and cleanup. Beverage producers use it for mixing, fermentation, cooking, cooling, and bottling. All too often the effluent is disposed of untreated into nearby soils or water bodies. Though there are few contami-nants and often plenty of nutrients, as untreated waste it can be toxic to aquatic ecosystems. Biochar can be used to harvest these nutrients while recovering the water for reuse, and then the nutrient-laden char can be used to grow crops, regrow forests, or regreen deserts.

A filtration-to-fertilizer (F2F) strategy has been explored in a joint research project of the Ithaka Institute for Carbon Intelligence, Rochester Institute of Technology, and Cornell University. Various biochars have been tested to filter wastewater from tofu, beer, and dairy whey. Early results show biochar reduces suspended solids and chemical oxygen demand while improving the pH in the effluents. Greenhouse trials using biochar filled with the nutrients from tofu production effluent showed that adding only 5 percent to potting mixes sig-nificantly boosted yields: 38 percent for lettuce and 62 percent for basil.

Currently many tofu, cheese, and yogurt manufacturers send tanks full of highly acidic whey to nearby farmers who spray it on fields as a low-cost disposal strategy. Let's reimagine that harmful scenario transformed by a beneficial carbon cascade: Farmers lease carbonized crop residues to these manufacturers, then recover the saturated and enriched char to use as a slow-release fertilizer. This could lower the cost of water and wastewater treatment for the tofu, cheese, or yogurt maker, boost the bottom line for farmers, and prevent soil acidification.

This type of cascade can displace what is often done with activated car-bon used for filtration. Because of its high cost, saturated activated carbon is often shipped off to be reactivated using high-temperature thermal processes that purge residues but burn off some of the original carbon at the same time. While regeneration creates fewer greenhouse gases than making activated carbon from scratch, the F2F approach eliminates the need for reactivation energy, reduces transportation miles, and reduces the need for fertilizer, which carries its own heavy carbon footprint.

All of this is well and good, you might say, but it sounds a bit like fairy dust. Maybe practical in small, isolated pockets of society, but is it really workable to address a crisis of planetary proportions? How do we avoid the dangerous consequences that characterize so many negative emissions technologies? Could it ever be economically viable? And where do we even begin?

Our first priority should be to source biomass from existing waste streams, particularly those that if neglected would generate greenhouse gases by being either burned or left to decay into methane and trace gases. Fortunately, there is plenty of waste biomass that could and should be carbonized. That orphaned resource represents energy, which means economy, and that can mean happiness for waste generators, processors, shareholders, and society as a whole. A straight-line economy begins with a valuable resource and ends with toxic pollution that is incapable of being recycled. That business model needs to be replaced by a vibrant circular economy where there is no such thing as waste. Waste is not a noun. It is a verb. We waste opportunities every time we waste.

Apart from bacteria, the total live biomass on Earth is about 560 billion tons of carbon. The total annual primary production of biomass, wild and domesticated, is just over 100 billion tons of carbon per year. Of that, farmed annual cereal crops are about 2.3 billion tons. And more than half the cereal biomass by weight is considered "waste." When we consider available "food-grade" wastes that could be turned into biomass energy, biofertilizers, pharmaceuticals, and other uses, there is ample supply waiting to be tapped to reverse the carbon cycle and begin drawing down legacy emissions.

Cereal crop by-products can go through several transformations — mashed for leaf protein extraction, fed to cattle or fish, fermented and distilled, dried for barn fodder, and carbonized for energy production — before returning to the soil to support new crop growth as compost and biochar. Ultimately rewarded in this way, soils are rejuvenated, robust, resilient, and ready to provide again for future generations. If we look at paper mill waste, poultry litter, hog farm slurries, mountains of coffee grounds, or any of scores of present-day pollution nightmares, we quickly discover how easy it is to pyrolyze those materials while reducing waste management costs and greenhouse emissions, leading to both economic and environmental benefits. Most importantly, the potential for drawing down atmospheric carbon jumps by orders of magnitude. Only through a change in perspective can we change the way our economy works — from a dysfunctional, marginalizing, and destructive Economy with a capital E to many smaller economies that build social, natural, financial, political, and cultural capital. Circularity means designing products that can be "made to be made again." And we can power that new system with energy that works the same way. In other words, we can cascade more than carbon.

Poudrette, a euphemism the Renaissance French applied to a fertilizer made from humanure, is perhaps one of the best examples of circularity. Millennia past, indigenous peoples combined charcoal with human feces to eliminate odors and speed composting. Returning human wastes to the fields was a foundational practice for sustaining agricultural productivity for more than forty centuries in China, Korea, and Japan. As sanitary standards evolved in cities, the contents of "dry closets" (as opposed to "water closets" that flowed into cesspools and sewers and thence to rivers), often referred to as night soil, were emptied and their contents were hauled to the outskirts and mixed with ashes, peat, gypsum, clay, lime, and more charcoal. Composting helps kill most pathogens. Adding biochar speeds the composting process and can increase temperatures within the compost pile, thereby killing off more pathogens. It can also subdue the stench.

In addition to closing the loop on carbon resources, we also need more forests, all over the planet, to draw carbon from the atmosphere and oceans and return it to where it belongs. There's no way around this. Photosynthesis is the suction for the carbon cycle and forests photosynthesize far more than fields and croplands. These forests must be managed, at least in the near term, to optimize growth and sequestration, a process that will gainfully employ hundreds of millions of us. China had the right idea when it recently assigned sixty thousand soldiers to plant trees over an area the size of Ireland. Norway, Sweden, and Finland recently crossed the balance line, thanks to good management practices, between forestry being a net carbon source and becoming a large carbon sink. The most important change that occurred in all three countries — Finland in 2005, Sweden in 2010, and Norway in 2015 — was a shift in forestry products from short-lived, single-use products like fuel pellets, pencils, and paper to long-term sequestrations in furniture, boats, and buildings.[3]

To be able to sustain itself in good times and bad, broadscale forestry must consider the social, economic, and environmental dimensions of each enterprise, including: food competition, basic services, family involvement, equal opportunities, child labor, access to land and land tenure rights, retirement benefits, regulatory regimes, scientific and technological innovation, self-financing, income diversification, soil health, chemical safety, net land use degradation, biodiversity, waste management, availability and reuse of water, training, family health care, children's education, lateral organization, and open, transparent participation. We prefer the word *ecoforestry* to convey all that.

Some speculate that the wood demand for bioenergy may rise so rapidly it could threaten existing forests, especially where regulations are lax or nonexistent. Several countries have responded with strict criteria, in some cases allowing only biomass from certified sources. The notion, put forward by opponents of biomass energy, that scaling up biomass to biochar with combined heat and electricity could lead to deforestation makes sense only if you ignore the amount of land, time, labor, transportation, and money that would be required, and also ignore the abundant biomass closer to home and available for the taking; in some cases, waste generators will even pay you to take it. Given the enormity of the available waste resource, we think the idea that biochar producers would cut down healthy trees and forests to make biochar is a straw man fallacy.

We agree: It is neither desirable nor practical to carpet the world with large-scale biomass energy farms. From the standpoint of ecosystem services and biodiversity both above- and belowground, not to mention the human social impacts, that sort of industrial bioenergy model would be a disaster. We also believe bioenergy should be primarily a side-product of forest harvesting and food and wood processing, and should be inherently a local enterprise, located close to the production of the excess organics. Planting and maintaining healthy forests is essential, and carbonizing thinnings or diseased trees is synergistic with that enterprise. The combination is better than either alone. "Photosynthesize and carbonize" ought to be the clarion call of the twenty-first century.

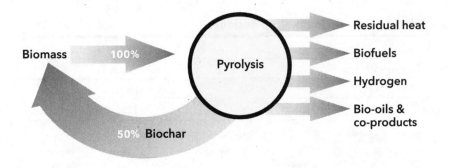

In the late 1970s, scientists at Oak Ridge National Laboratory proposed a model to make profits while making biochar. The model takes biomass from sources that do not compete for available farmland and emerges in two parts: half as biochar that returns to the soil and half as other commercial products and services.

Until now, biomass energy crops have intensified the negative impacts of the agricultural sector on ecosystems, manifested as deforestation and forest degradation, loss of biodiversity, soil degradation and erosion, and water depletion and pollution. Consequently, governmental and nongovernmental institutions, educational institutions, and society in general have genuine concerns about the sustainability of using land to grow crops for energy. These are well-merited concerns, but carbonization is not the same as combustion. Using healthy trees for biochar is not even economically viable compared with using other underutilized sources.

And yet, we feel it's important to acknowledge that energy is the heartbeat of any economy. As the residents of Puerto Rico can attest after Hurricane Maria, when the energy flow slows, the heart of a community slows, activities come to a halt, and the economy withers. Unless we can morph quickly toward a combination of renewable energy, vastly more efficient energy use, and carbon-removal technologies, we'll soon experience this far beyond the shores of Puerto Rico. Carbon cascades can get us where we need to go while keeping that heart beating.

In the late 1970s, scientists at Oak Ridge National Laboratory proposed a model to make profits while making biochar.[4] The model takes biomass from sources that do not compete for available farmland and emerges in two parts: half as biochar that returns to the soil and half as commercial products and services. Until now, the focus has remained on using biochar as a soil fertility

Table 4.1. Carbon Math: Crop Residues

Total annual crop residues*	4.0 GtC/yr
Convert 25% to biochar	1.0 Gt/yr
With 50% C content	0.5 GtC/yr
CO_2-e drawdown potential	(1.8 $GtCO_2$-e/yr)

Note: Potentials for atmospheric removal are in parentheses.

* Qingzhong Zhang, Zhengli Yang, and Wenliang Wu, "Role of Crop Residue Management in Sustainable Agricultural Development in the North China Plain," *Journal of Sustainable Agriculture* 32, no. 1 (2008): 137-148.

booster with the (usually nonmonetary) bonus of being able to safely sequester carbon for the long term. In much of the world, markets for sequestration are nearly nonexistent, and some biochar producers are stuck in a quagmire of low demand due largely to low market awareness while biochar users are forced to pay unreasonably high prices due to limited production. The current price for a delivered ton or even a small sack of biochar, either raw or blended, is still relatively high, and that makes the business case a tough sell if you are thinking of it merely as something to increase the yield of your carrots or corn. The math just doesn't pencil out for price versus profit.

Take maize and soybeans. They cover a huge area of the best farmland in North and South America. Even in a good year, these crops will gross less than $1,000 per acre ($2,470/hectare). In 2016, average gross profit of maize was $762 per acre ($1,882/hectare) and $653 per acre ($1,613/hectare) for soy.[5] Adding even 1 ton of biochar would turn profits into losses. To break even, assuming you were able to find biochar at $600 per ton, you'd need to nearly double your yields. It's unlikely 1 ton per acre would do that unless your soils were extraordinarily poor. While the economics are different for coffee growers in Africa or almond growers in the Middle East, for most annual cereal crops in good soils the bottom line is: Biochar is not cost-effective at current market rates.

As we have noted, this focus on soil is unnecessarily restrictive when biochar, active charcoal, pyrolysates, or whatever new or old name is given to the material has so much more to offer in terms of both climate change abatement and environmental regeneration. As we saw, the most interesting characteristic of carbon is not its potential to improve soil fertility but

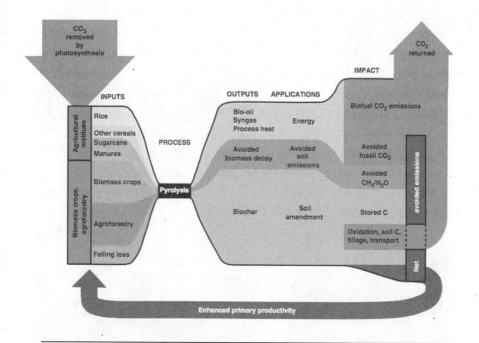

In this model cited by Pete Smith, the potential carbon sequestration yield from biochar is limited to crops and residues, and the applications are limited to agricultural uses.

its nearly unmatched versatility. It can recycle nutrients and upcycle waste. It can remove pollutants and retain water. It can immobilize and catalyze. It can multitask. It can have cascading, sequential lives.

At a webinar hosted by the US National Academies of Sciences, Engineering, and Medicine in September 2017, Pete Smith, professor of soils and global change at the University of Aberdeen, Scotland, carefully tallied all the potentials for carbon sequestration by various means and determined that while technically it might be possible to make 1.8 gigatons of biochar every year, competition for feedstocks and the ability of soils to keep receiving carbon limits the potential to 700 million tons per year. Smith showed a slide from a 2010 study.[6]

The figure shows inputs, process, outputs, applications, and impacts on global climate. Within each of these categories, the relative proportions of the components are approximated by the height and width of the various fields. Starting at the point where CO_2 is removed from the atmosphere by

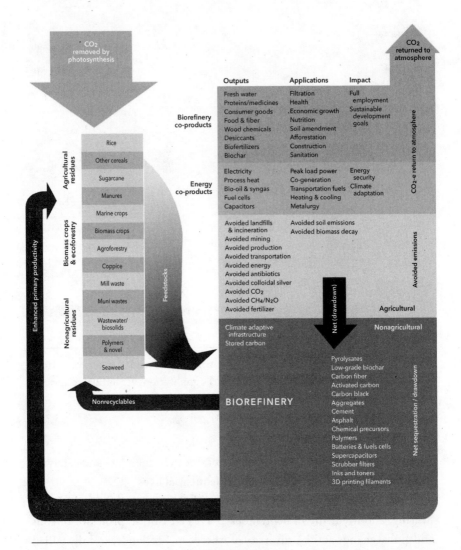

Our 2018 revision of the model includes pyrolysates unsuitable as soil amendments but which can be used for nonsoil applications. In this model, the potential carbon sequestration is more than 50 $GtCO_2$-e/yr.

photosynthesis, a sustainable fraction, such as agricultural residues, biomass crops, and agroforestry products, is converted to bio-oil, syngas, process heat, and biochar. The bio-oil and syngas are subsequently combusted to offset fossil fuels while the biochar does more than offset: It removes carbon from the cycle.

This was the state of the art in 2010. But science never sleeps. As we surveyed the large volume of research in preparation for this book, we felt the need to revise that chart to better reflect the expanding field.

With this new perspective, a world of opportunity opens in terms of how and how much biochar can displace materials that either have a large carbon footprint, are nonrenewable, are toxic, or are simply too expensive. By carbonizing underutilized waste streams and customizing production facilities, biochar can go a long, long way toward helping individuals, companies, and nations to meet their decarbonization targets. And, unlike other negative emissions technologies, these carbon-centric biorefineries also tick many of the boxes for the UN's Sustainable Development Goals.

After the Paris Agreement was signed, there began an accelerating discussion on the economic potential of decarbonization. This has educated and incentivized both investors and inventors. Our proposal is not only to broaden the focus of using biochar as a fertilizer carrier, but to include new, previously unimagined products and services that may provide higher returns on smaller volumes of carbon compared with agricultural uses. Some, like 3D printing and supercapacitors, may take time to develop but show promise in very lucrative markets. And, as we saw above, there is also enormous potential for carbon to create a new, bio-based, circular economy. Asking "What else can carbon displace?" is likely to lead us to products and markets we haven't even begun to consider. Asking "How much *could* a city or an industry or a country reduce its carbon footprint by carbonizing *all* underutilized organics and utilizing *all* by-products of pyrolysis or gasification?" could lead us to amazing new closed-loop industrial design scenarios and an economy where nothing is wasted and everything is upcycled.[7]

Imagine if we could recapture the enormous waste implicit in a subsidized, fossil fuel–based, linear economy — one that moves from extraction to processing consumption, waste, and pollution — by closing the circle and cascading the wastes back into products and services. With carbon cascades we are actually trying to break, or at least significantly slow down, one of nature's most hardwired cycles: the carbon cycle. But by applying the concept of circular economies, by repeatedly using carbon in various ways before it is entombed underfoot, we hope to avoid the economic and environmental fiascoes we now find ourselves in due to our frivolous use of fossil fuels.

Biomass residues transformed by fire are no longer the exclusive domain of game rangers, grain growers, soil scientists, agronomists, plant nutritionists, and plant biologists. Pyrolytic conversion is starting to engage the material sciences, electrochemistry, civil engineering, biophysical economics, electrical engineering, veterinary sciences, waste management, climate science, and public policy. This is only the start. As the full impact of climate change descends upon us, we'll need to move beyond academic research silos and beyond patent-driven, closed-door competition. We'll need collaborative, open-source teams working together across borders and across disciplines to develop carbon-sequestering products and processes.

Companies eager to cash in on the drawdown market and attract millennials with no desire to repeat the linear economy mistakes must build a more ambitious model — one that reimagines this orphaned biomass. They can design innovative cascades of foods, feeds, medicines, fibers, biocomposites, water filtration, electricity, process heat, bricks, roads, and bridges before the same carbon material, born of sunlight, ends up in soils to restore, reinvigorate, and regenerate planet-saving ecosystems from droughts, floods, hurricanes, and plagues of locusts. This is an industrial third wave, the carbon revolution that can provide employment that cannot be off-shored and can be launched without the taxpayer subsidies that have enabled oil and gas companies to stay in business for far too long.

This new way of thinking offers a way for urban environments to begin a transition toward negative emissions while reducing costs associated with what we incorrectly call waste management, but is really organic mismanagement — the squandering of valuable carbon. Communities spared from hosting a landfill in their backyard will also benefit. When biochar is used as plaster, paints, wallboard, roofing, insulation, or bricks, buildings can still be recycled after some decades to hundreds of years — becoming soil conditioners or new building materials. This is a centuries-long carbon cascade.

———

Human economics, like humans themselves, evolved in an era of nearly unfathomable natural abundance. To classical economics, however, nature's abundance is a neglected externality. It is seldom considered. It is just always there. Some peak oil theorists imagined twenty years ago that climate change would suddenly abate when we ran out of economically extractable fossil reserves. We passed that point for conventional sources in January 2011 (at 86.2 million barrels per day), but kept going, moving into the domain of more expensive, still marginally profitable, unconventional sources, such as oil shale, tar sands, and fracked gas. Those unconventional sources, along with what remains of coal and sweet crude, are crossing into the "unaffordable" column, in part because the Paris Agreement put them on death watch, and in part because they are, one by one, becoming money losers for those who mine and drill.[8]

Carbon cascade entrepreneurs are extraordinarily lucky right now. For better and for worse, they have time to prepare — time to plan, build prototypes, test their technologies, products, and business models. This extra time has been paid for by the coal magnates and giant oil companies — those who spent billions of dollars to sow climate confusion and, failing that, purchase politicians at the highest levels of government. While the delays have been disastrous for polar bears and coral reefs, and unconscionable toward those affected by natural disasters of biblical proportion, it has purchased a little time for solutions to be more carefully vetted and scaled. We will look at some of these solutions more closely in the two parts to follow.

PART II

Carbon Construction: A Fresh Foundation

Mr. McGuire: I just want to say one word to you. Just one word.
Benjamin: Yes, sir.
Mr. McGuire: Are you listening?
Benjamin: Yes, I am.
Mr. McGuire: Plastics.
Benjamin: Exactly how do you mean?
Mr. McGuire: There's a great future in plastics. Think about it. Will
you think about it?

— THE GRADUATE *(1967)*

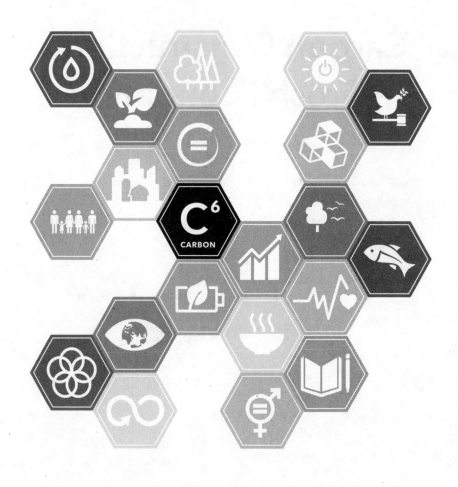

CHAPTER 5

Carbon Hardscaping

Things move along so rapidly nowadays that people saying: "It can't be done," are always being interrupted by somebody doing it.

— Puck MAGAZINE *(December 1902)*

oncrete is the world's second most consumed material after water. Owing to rapid urbanization, especially in Asia, global demand for construction aggregates, the largest part of concrete, is projected to reach 66.3 billion tons by 2022, an $84 billion market.[1]

A necessary ingredient of concrete today is a binding agent made from pulverized limestone (calcium oxide) and clay (aluminum silicon oxides) heated at high temperature (1482°C/2700°F) to make Portland cement. The discovery and refinement of Portland — now at nearly 5 billion tons per year and said to be responsible for 5–8 percent of global greenhouse pollution — is a cautionary modern tale of the intersection of materials and manufacturing from the dawn of the fossil fuel era.

The Romans and Chinese millennia ago discovered gypsum and lime could be mixed with pieces of rock, sand, clay ceramics, or rubble to form a hard material that would hold up to weather, or even set up underwater to construct dams and bridges. Roman concrete, developed from 150 BCE, is durable due to its incorporation of volcanic ash and cinders (pozzolana), calcined clay, and aged lime, which prevent cracks from spreading. After the famous fire of AD 64, Nero rebuilt much of Rome with brick-faced concrete. The Pantheon in Rome, with its 142-foot coffered dome and oculus, still stands after two thousand years, a testament to the durability of concrete.

When limestone is kilned to make lime for mortar or cement, 1.8 tons of stone produces 1 ton of CaO (calcium oxide or quicklime). The missing 0.8 goes to the atmosphere as CO_2 and trace impurities. China is by far the world's largest producer of cement, burning enough rock to produce around 170 million tons per year. The United States is the next largest, with around 20 million tons. Worldwide, lime kilns send about 225 million tons of CO_2 to the atmosphere each year. File that number away for a moment.

Quicklime is a white, caustic, alkaline, crystalline solid at room temperature, but feeling an urge to go back to rock, it will draw CO_2 from the air unless slaked with water. Slaked lime is what the Romans and Chinese used for mortars and plasters. It is what Michelangelo in 1511 spread across the ceiling of the Sistine Chapel and painted the image of God into. Nineteenth-century vaudevillians learned that when heated to over 2200°C (4000°F) it emits an intense glow, which was useful in theater productions before electric lighting — limelight.

If you add an atom of carbon to quicklime in the presence of oxygen, you get limestone and water. That limestone molecule can take a much-deserved rest. It has now completed a full revolution on the wheel of life.

Charoset

A disruptive technology that may someday make Portland obsolete owes much to the research of an Australian, Dean Farago, who traced the journeys of his own family history back seventy-five hundred years to when the indigenous peoples of the Levant learned to mix two of their most abundant and otherwise worthless materials — clay and ash — to make cement.

Farago said it would be common when building a house to start by finding the best site for an earthen cistern to store water. "The gasification kiln was usually made/moved so as to bake the wall of the cistern" while making char, ash, and calcined clay. The clay came from the pit dug where the cistern, and then the house, were to be built.

"Nothing was wasted and that especially included energy," Farago told us. The char would be used for either cistern filter material and replaced on occasion with fresh char, or as a nonreactive lightweight aggregate for the charoset. To Farago, traditional charoset was made of calcined clay (750°C/1382°F for four hours), potassium lye, and an aggregate, which could be sand, gravel, or biochar. For more flexural strength you would add a silica ash, derived from burning grasses, palm fronds, or bamboo.

The system he described for the house cistern was an *ahavirot*, or trompe pump (water-powered air compressor), that used a difference in gradient for falling water to generate compressed air. "The original purpose of it was to solve the problem of stored water becoming anaerobic below 3.5–4 meters (11.5–13 feet), depending on altitude and climate," he said, adding that compressed air had more uses than calcining clay and aerating water.

"Then the use of compressed air to condense water from the air, especially in high-surface-area plasterwork, was worked out. This then allowed for yet more water to be directed down columns and into cisterns, producing yet more compressed air. Refrigeration and freezing came after. . . . It's also part of how their forestry orchards were planned . . . as the excess compressed air could be directed towards airwells along the orchard rows that allowed the cold fog/mists/frosts coming down to be directed towards the apples/pistachios — the trees that need a little frost — and away from the citrus, that didn't want it. It looked like magic when they first showed me when I was a kid.

"Airwells are another part that are integral to the way I was taught to build. The 'decorative' plasterwork, especially when made colder from the mortar mix and from 'waste' compressed air, can be a huge part of harvesting water, and also recycling water from the air, as in courtyard gardens that were very popular in the Mediterranean. My grandparents' home had decorative sebkha plasterwork on the exterior of the fridge/freezer room that would collect 300–700 liters (80–185 gallons) per day."

Farago reminded us of the Australian farmer Percival Alfred Yeomans, who developed a system for managing water in a dry climate he called "keyline" management. Yeomans's 1973 book, *Water for Every Farm*, is still in print. Farago said the ancient system designed to resemble a flower was really a keyline, and that keyline had always been a landscape management/building/water system. "Imagine it like a honeycomb design over the landscape running towards the ridges. This would decide where you built your cisterns, then your home, and where to plant what."

The composition of the mortars was very important. Pulverized charcoal would be reserved for these purposes. Mix the right combination of an aluminum silicate (calcined kaolin clay), an aged lime, and an aggregate (biochar, sand, sawdust, or gravel fines), and you get a concrete that can be lightweight and insulating and yet fast-setting, and that becomes as strong and as monolithic as marble or granite.

Farago told us he "came up with a floating platform with a ballast for an *ahavirot*. It's great for coastal areas or dams that have gone anaerobic." He said he'd like to advance to an ocean platform for pulling out plastic and processing it. "With the compressed air we could clean up the oceans and oxygenate them again. It would deacidify them a fair bit as well.

"I would advise anyone with a woodstove or oven to, from now on, collect the ash and store it in a dry, watertight plastic bucket, and this includes palm fronds as they are especially valuable, and treat this ash like gold . . . and also, to weld up a metal tray that each time you have a fire . . . you take dry clay that has been crushed into a powder and calcine this in your fire. And collect this and also keep it dry. In biblical times people paid their taxes this way."

To ensure a concrete structure will last, inspectors test the strength of the concrete after twenty-eight days of drying. The compressive strength is rated using megapascals (MPa). The MPa of concrete ranges from 17 in homes to 28 in commercial projects. Any concrete product with an MPa over 41 is considered "high strength," and is good for large-scale projects like bridges or high-rise buildings.

"Depending on the effort and technique you will easily achieve a 20–60 MPa concrete, and depending on the aggregate, you will have a stone," Farago told us. "It's by far the most durable and strongest mortar method that I know of. I've made samples that reached 217 MPa [31,465 psi]."

Making Sand

Concrete today relies on enormous amounts of another resource that once seemed in endless supply, but is disappearing rapidly: construction sand. Sand is still priced by the cost of extraction, not on replacement value, leading to rapid depletion. Despite this, the price at the mine in the United States has barely changed over the past century.

As odd as it may sound, in the Middle East, they import sand for building purposes from Australia. Sand from deserts is not desirable for most construction needs, since wind polishes grains until they do not bind well and have low shear strength. Sharp-edged sand grains with a rough surface are needed to bind mortars.

The annual world consumption of sand is estimated to be 15 billion tons, worth $70 billion. Sand has by now become the most widely

How in the World Can We Be Running Out of Sand?

Riverbeds, oceans, and deserts are full of sand, yet certain types of sand are being depleted so rapidly that some countries are putting bans on exporting them. India consumes 500 million tons of sand annually, but that's only the legally recorded amount. The black market is estimated at $16 to $17 million *per month*. Nicknamed "red gold," India's sand has larger individual particles, and its superior compressive strength makes it highly prized for cement construction and glassmaking.

After twenty-six-year-old Brijmohan Yadav reported the mustard-yellow excavator machines digging his family's land, uprooting crops to create a road to the riverbanks, and wrecking the small embankments he had built to catch the rainwater his family needed for farming, he says he has been intimidated and attacked by enforcers of the illicit trade. A gang of men went to his house and threatened to kill him if he did not withdraw his complaint. Now he is in hiding.

"I am in danger . . . my family is in danger," he told a reporter over the phone from an undisclosed location. "The sand mafia is very strong."[2]

consumed natural resource on the planet after water, and concrete is the biggest consumer. One kilometer of road requires 30,000 tons of sand. A single house might contain 200 tons. Chinese demand rocketed 437.5 percent over the past twenty years while demand in the rest of the world increased by 59.8 percent.[3] Industrial sand consumption is twice the yearly amount of sediment carried by all of the rivers of the world.

Since 2005, at least twenty-four small Pacific islands disappeared as a result of illegal sand mining. Most of it was shipped to Singapore, which has expanded its surface area by 22 percent since the 1960s. Sand removal is causing declines in seagrasses in Indonesia and threatening extinction of Ganges River dolphin and gharial (a rare crocodile), terrapins in India and Malaysia, and migratory birds in eastern China.

Table 5.1. Carbon Math: Sand Displacement

Global sand mining	15 Gt
Replace one third with biochar	5 Gt
With 82% C content	4.1 GtC/yr
CO_2-e drawdown potential	(15 $GtCO_2$-e/yr)
CO_2-e avoided from sand mining and transport	unknown

Note: Potentials for atmospheric removal are in parentheses. The degree of substitution is based upon the physical properties of the substance being displaced and the authors' estimate of market penetration potential.

The extent to which biochar can displace sand and gravel as a construction aggregate depends in part on the end use of the concrete, but it has been trialed at up to 30 percent for non-load-bearing applications. Because the bulk density of biochar is less than sand, this lightens the weight considerably, but the higher porosity of biochar can also increase the amount of water needed to keep the workability of wet concrete in good order as it is poured.

But here is the kicker: Carbon reacts differently than silicon when used this way. Depending on the amount and type of biochar used, carbon concretes and mortars may improve:

- weight (biochar is significantly lighter than sand)
- compression strength
- flexural strength (MOR)
- curing time
- capacity to absorb CO_2 and nitrogen oxide
- electromagnetic shielding
- fire resistance
- insulation
- humidity control
- indoor pollutant control (dust, pollen, chemicals)

Biochar is being tested and used in concrete recipes, but so far the motivation has been either carbon sequestration or the production of lighter concrete. Based on cost alone, biochar can't compete; the price of sand has

Table 5.2. Carbon Math: Construction Aggregates Displacement

Global construction aggregates (excluding sand)	53.0 Gt/yr
Convert 20% to biochar	10.6 Gt/yr
With 82% C content	8.7 GtC/yr
CO_2-e drawdown potential	(31.9 GtCO$_2$-e/yr)
CO_2-e avoided from aggregate mining and transport	unknown

Note: Potentials for atmospheric removal are in parentheses. The degree of substitution is based upon the physical properties of the substance being displaced and the authors' estimate of market penetration potential.

been escalating due to increased demand, but still costs less than $10 per ton, and biochar sells for between $500 and $2,000 per ton, though gradually the price is falling as new production comes online.[4]

When the improvements biochar brings to concrete are factored in, the value proposition begins to change. We know that biochar added to cement can help reduce cracking and improve flexural strength as compared with using just sand for the fine aggregate. We know from a study appearing in the *Journal of the Korea Institute for Structural Maintenance Inspection* that chemical and mechanical properties — components, microstructure, concrete weight loss, compressive strength, and mortar flow — for biochar are as good as or better than for other aggregates, including fly ash, and that certain types of biochar "reduce water evaporation from concrete which reduces both the plastic shrinkage and drying shrinkage." Carbonized biomass could potentially reduce liabilities related to concrete failure, reduce curing time (which means faster building), and provide better insulation (which reduces building operating costs).[5] Engineers suspected, and soon confirmed, that adding small amounts of very fine particles of high-grade biochar — for example, sawdust biochar, created at 500°C (932°F), yielding 87 percent carbon — at a rate of 2 percent by weight of the cement, produces a stronger concrete.[6]

Consider the carbon math for just a 20 percent market penetration.

Recent studies in Italy and Switzerland suggest that biochar concrete ("charcrete") may also provide fire resistance, especially in high-temperature tunnel fires.[7] When concrete is exposed to fire above 1200°C (2192°F), its embedded water is turned into vapor that has no escape and builds up large tensile stresses within the cement-aggregate matrix. When such tensile stresses overcome the tensile strength, concrete will explode violently (spalling), causing damage to the structure and potentially injuring firefighters and rescuers. The usual solution to this problem is to use polypropylene fibers, typically 2–3 kg/m^3 dispersed through the concrete mix. Unfortunately, these fibers partially melt at about 250°C (482°F), leaving behind a network of interconnected capillary pores allowing water vapor to build.

Enter biochar. Like polypropylene, biochar fibers act as disturbing elements, and a network of capillary microcracks develops within the matrix. During fire exposure biochar relieves the internal pressure by adsorbing water vapor through the capillary microcracks. This provides, in combination with suitable cementitious materials, spalling resistance for concrete exposed to fire in road and railway tunnels. Along with traditional ventilation, it may also help lower pollutant levels in road tunnels.[8]

Another feature of these water-filled capillary microcracks is that biochar becomes a carrier for bacterial spores in the cement.[9] These embedded spores give concrete the extraordinary ability to self-heal hairline cracks and small fissures and reduce permeability within concrete so water does not accumulate. The microbes digest minerals like calcites and silicates and exude semicrystalline glues that arrest fissure propagation. Given that the cost of repairing concrete can often be higher than the original cost of pouring it, this remarkable attribute holds enormous economic potential.

But wait! Biochar's porosity adds another benefit: low thermal conductivity. The entrapped air within the pore spaces can improve concrete's normally poor ability to insulate. The demand for electric power to run air-conditioning in hot climates is increasing as climate change accelerates. Added power has become essential for cooling purposes and comfort inside buildings in a steadily growing number of regions where outdoor temperatures in summer can reach up to 60°C (140°F). Use of insulating materials is often unpopular, despite their long-term financial benefit, because it can add to the cost of building. However, merely changing the standards for building materials used in new construction, such as the use of composite blocks with high thermal insulation properties, may provide cascading carbon benefits. The development of lightweight concrete blocks for thermal insulation is already under way.[10] Passive air-conditioning of the sort theorized by

> We believe ultra-lightweight concrete is one of the most
> fundamental bulk materials of the future.
> — CHRISTOPHER ALEXANDER et al.,
> *A Pattern Language*[11]

R. Buckminster Fuller in the 1940s to chill air drawn by a vacuum effect, or by Dean Farago's ancestors seventy-five hundred years ago to make chilled compressed air for refrigeration, is likely to make a comeback.

One of the more exciting aspects of our newfound ability to substitute carbon for silicate in high-rise buildings and other hardscape is that it opens up a vastly larger range of potential feedstocks that could be pyrolyzed. If we were confining our pyrolytic carbon product to use in soil, we would need to be mindful of heavy metals and other long-lasting toxins that should be prevented from entering terrestrial or marine food chains. If we instead entomb these toxins in buildings and roads, bridges and tunnels, and seal them properly, neither the carbon nor contaminants will go back into the environment. Ever.

Well, ever is a long time. Maybe we should say, not on the time scale of millions of years.

Different types of biochar, such as from hazelnut shells, coffee residues, poultry litter, rice husks, or pulp and paper mill sludge, have been tested in varying amounts for use in concrete and mortars with differing results.[12] There are a lot of variables that influence success or failure, both with the concrete/mortar and with the biochar. We are still in the early stages of understanding which biochar properties are most relevant to cement, but particle size, porosity, and water-holding capacity appear to be key considerations. It may be that silica content, hydrophobicity, and other characteristics are also important in particular applications.

By testing carbon concretes at 5, 10, 15, and 20 percent biochar content by volume, researchers learned:[13]

- All biochar admixtures had less weight loss due to moisture evaporation. Mortar mixes with biochar have better water retention, and this may lead to improved strength.

- Although the internal water content is higher, workability of mortar decreases as the percentage of biochar increases because it dries faster.
- An addition of 5–10 percent biochar performs similarly to 20 percent replacement with fly ash, the toxic residue of power plants and other industries.
- Adding up to 5 percent biochar shows an increase in compression strength. At between 5 percent and 15 percent by volume, additional materials such as charoset are needed to maintain the compression strength.
- Carbon cement outperforms silica-based controls in bending strength, compression, and fracture energy.
- Biochar from coffee bean discards performs better on compression tests than other types of biochar that were tested or than cements without it.
- Biochar from hazelnut shells performs better on flexural (MOR) and fracture energy tests.
- Hazelnut shells' irregular morphology, transferred to biochar, creates a "perfect bond with surrounding matrix."
- Coffee powder–based biochar has higher silicates that could work as an accelerator, helping speed up the hydration process. It stabilized at seven days.

Concrete engineers have found in repeated tests worldwide that using similar amounts and types of biochar, and presaturating biochar six hours prior to mixing, adds higher compressive, flexural, and split-tensile strength.[14]

There is a large potential for rebalancing atmospheric chemistry by changing the way we make cement. We know that the world produces 4.2 billion tons each year of Portland cement by heating limestone and clay minerals in a kiln to form clinker (a cement-making by-product ground to make mortars), grinding the clinker, and adding 2–3 percent of gypsum. Mining the limestone, clay, and gypsum, heating the kilns, grinding, and transporting all take energy, and that life-cycle cost can be given in carbon dioxide terms: 9 pounds of CO_2-e for every 10 pounds of cement.[15]

If we replace half of all Portland cement with charoset, we still have some life-cycle costs. Although the ash and pyrolysates (assuming 10

Table 5.3. Carbon Math: Cement Displacement

Global cement production by weight [2017]	4.2 Gt/yr
Displace 1% with biochar/charoset	0.04 GtC/yr
With 82% C content	0.03 GtC/yr
CO_2-e drawdown potential	(0.15 $GtCO_2$-e/yr)
CO_2-e emissions from cement production	3.8 $GtCO_2$-e/yr
Avoided CO_2-e emissions	0.1 $GtCO_2$-e/yr

Note: Potentials for atmospheric removal are in parentheses. The degree of substitution is based upon the physical properties of the substance being displaced and the authors' estimate of market penetration potential.

percent by weight would be optimal) are waste products, we have to mine and calcine the clay. Still, eliminating limestone takes 5 of those 9 pounds of CO_2-e out of the process, dropping the life-cycle impact by 55 percent. If we take our pyrolysates from waste sources, they may contain 50 percent carbon (versus 82 percent for biochar) or less, but whatever carbon content is in that cement is carbon that will (almost) never go back to the atmosphere.

If 1 percent biochar or carbon pyrolysate by weight were added as binders to each year's 4.2 billion tons of cement and 3.7 billion tons of clinkers, 40 million tons of carbonized material would be entombed annually. With a carbon content of 50 percent (for pyrolysates) to 82 percent (for biochar), the amount of actual carbon entombed in cement and mortar could be 20–30 million tons per year, apart from the aggregates (sand and gravel). Converting that to CO_2 equivalent, we could be pulling more than 100 million tons out of the carbon cycle. Now imagine 10 percent, then 20 percent.

Adding biochar to cement, while not a silver bullet, is yet another silver pellet in a buckshot strategy of carbon drawdown. Switching 10 percent of our cements and mortars from silicates to carbon would cancel out another 1 percent of annual greenhouse gas emissions. At the end of the day, even replacing 50 percent of cement worldwide with charoset wouldn't give us a clean bill of climate health. We would still be adding 2 gigatons of carbon dioxide to the atmosphere to make the remaining cement. That is not as bad as the current cement-driven 3.8 gigatons going to the sky each year, but it doesn't escape the predicament.

Roofing Tile

With students from Rochester Institute of Technology, Kathleen tested roof tiles made from a lightweight aggregate of biochar, sand, shredded plastic bottles, and cement. The method was simple: Mix all the dry ingredients, add water, blend thoroughly, pour into flat tile mold, vibrate to remove air bubbles (an electric sander works well), transfer the flat tile to curved tile mold using a plastic sheet, and allow two weeks to cure. The results:

- Each tile weighed around 14–16 pounds (6–7 kg).
- The tiles supported a 210-pound (95 kg) person standing on them.
- With two workers, it will take five days to make 224 tiles for one home.
- Estimated CO_2 saved per roof is ~400 pounds (182 kg).

The students discovered that an entire roof could be produced of durable, locally available biocomposites made of cement, recycled plastics, and biochar for less than $400. These lightweight roofs could withstand a hurricane, cut the noise (from rain) by 25 percent, and insulate against daytime heat 20 percent better than the previous metal roofs.

Does carbon concrete remove CO_2 from the atmosphere? Not directly, apart from the rebalancing of lime, but using feedstocks like peanut shells, hazelnut shells, and coffee bean discards that would otherwise return to the atmosphere in the form of carbon dioxide or methane interrupts emissions and holds the carbon in the soil or the built environment for a very long time. At sufficient scale, once we remove more from the input side of the atmospheric and oceanic carbon ledger than is naturally withdrawn on the output side — rock mineralization, peat formation, ocean sediment deposition, and so on — the balance will begin to slowly return to its preindustrial optimum.

At that point we can claim that there's nothing inherently wrong, from a climate perspective, with building airports, parking lots, bridges, tidal barriers, and roads, because we're sequestering carbon — concretely. Plus, reduced sand, reduced water-wicking, and self-repairing concrete mean reduced liabilities and reduced dependence on dwindling resources to the construction industry, and whole new lines of lower-carbon products to be rolled out. Carbon-based concretes are an entrepreneurial opportunity waiting to be seized.

Paving the Way

Almost fifty years since Joni Mitchell began singing "They paved paradise, and put up a parking lot," paving over paradise is still a booming business — more than $2.2 trillion is spent worldwide on new airports, roads, bridges, rail routes, ports, tunnels, and parking lots every year. An enormous amount of materials go into converting permeable surfaces into impermeable ones, including asphalt and concrete but mostly aggregates in the form of stones, sand, and fillers. Though concrete roads last longer, a majority of roads are surfaced with the lighter, more malleable asphalt.

Europe and North America have by far the most extensive networks of paved roads and highways in the world. In Europe, more than 90 percent of paved roads and highways (3.2 million miles/5.15 million km) are surfaced with asphalt. In the United States, it's over 92 percent (2.5 million miles/4 million km). In addition, about 85 percent of US airport runways and 85 percent of parking areas are surfaced with asphalt. Then add Canada (258,000 mi/415,211 km), Mexico (110,000 mi/117,028 km), Central and South America (176,000 mi/283,245 km), Australia and New Zealand (77,000 mi/123,919 km), China (979,000 mi/1.6 million km), and 1.3 million miles (2.1 million km) over the rest of Asia. To meet these needs, 1.8 billion tons of asphalt is poured every year.[16] By adding 2 percent biochar or bio-oils, that industry could be creating a demand for 32 million tons of biochar or pyrolysates each year.

Laura Ingalls Wilder, nineteenth-century author of the *Little House on the Prairie* series, tells of her first encounter with an asphalt pavement as her parents drove their covered wagon through Topeka.

> In the very midst of the city, the ground was covered by some
> dark stuff that silenced all the wheels and muffled the sound

of hoofs. It was like tar, but Papa was sure it was not tar, and it was something like rubber, but it could not be rubber because rubber cost too much. We saw ladies all in silks and carrying ruffled parasols, walking with their escorts across the street. Their heels dented the street, and while we watched, these dents slowly filled up and smoothed themselves out. It was as if that stuff were alive. It was like magic.

Bitumen, a thick, gooey by-product of petroleum processing, is used as a binder in asphalt construction. A typical recipe is 5 percent bitumen blended with 95 percent aggregates, with bitumen functioning as the glue that binds the mineral aggregates into a cohesive mix. Every asphalt pavement mix is designed for a specific pavement application, varying its composition accordingly.

Though asphalt is often chosen over concrete, it is susceptible to various environmental stresses and some of the fillers used — crushed rock, gravel fines, sand, and construction rubble — are not as well suited as newer replacements, including composites of recycled tires, glass, wax, elastomers, plastomers, adhesion promoters, anti-stripping agents, and industrial chemicals like phosphoric acid. This additives market was $3.3 billion in 2016 and is likely to rise to $4.7 billion by 2025.

Carbonaceous materials have long been used as asphalt additives because they are considered inherently compatible with bitumen, a hydrocarbon. This has motivated numerous studies since the 1960s to modify the asphalt binder with carbon-based materials for improved performance. Carbon fiber and carbon black are commonly used but both are expensive and come with a large greenhouse gas footprint compared with biochar or its pyrolysis companion, bio-oil.

In Malaysia researchers are looking at how carbonized organic wastes can improve the longevity of pavements.[17] Using up to 6 percent biochar — what Malaysians call *nanocharcoal* due to small particle size (less than 100 nanometers) — raised the temperature at which pavements soften. As climate warms, this will be important. Coconut shell biochar was found to improve the penetration or stiffness of the bitumen, which researchers theorize is due to the increased surface area for bonding. In Tennessee, researchers blended switchgrass biochar at 10 percent by weight with

Table 5.4. Carbon Math: Asphalt Displacement

Global asphalt production by weight [2017]	2.5 Gt/yr
Convert 20% by weight to bio-oils, biochar, or pyrolysates	0.5 Gt/yr
CO_2-e drawdown potential	(1.8 GtCO_2-e/yr)
CO_2-e avoided from aggregate mining	unknown

Note: Potentials for atmospheric removal are in parentheses. The degree of substitution is based upon the physical properties of the substance being displaced and the authors' estimate of market penetration potential.

asphalt and found that it reduced the temperature bending variability in asphalt binders.[18] The switchgrass biochar blend also showed the highest rutting resistance, meaning it needs to be replaced less often.

While these reasons for incorporating biochar as part of a bitumen mix are compelling on their own, biochar or other pyrolysates could displace a portion of the aggregates normally used in road and roof construction — representing a much greater volume of materials. On rooftops, weight reduction and crack resistance are advantages. Biochar and bio-oils could be useful along highways, in stabilized embankments, or in structured soils and ditches, where it would not only control erosion but detoxify stormwater and neutralize road salts and petroleum residues.

Industrial trade groups Asphalt Institute and Eurobitume estimate the current world production of bitumen is approximately 87 million long tons per year.[19] There are more than 250 known applications, but 85 percent are in pavements for roads, airports, and parking lots. Ten percent goes to roofing and the remainder is used for sound deadening, water pipe coating, paints, waterproofing, and sealing materials.

The rising cost of crude oil has affected the asphalt paving industry and encouraged research into alternative and sustainable binder materials that could be economically and environmentally viable for pavement. Of special interest is bio-oil through pyrolysis of timber waste, oil palm waste, rice husks, coconut trunk fibers, municipal waste, and sugarcane waste.[20] After fast pyrolysis, bio-oil still contains some water and oxygen. The carbon content is about 55–60 percent. With some optimization, a 40 percent or greater substitution of bio-oil for petroleum binders is feasible.

Let's run the numbers. If 20 percent bio-oil, biochar, or carbon pyroly-sate by volume were added as binder additives and aggregates to 2.5 billion tons of asphalt poured each year, 500 million tons of carbonized material would be entombed annually. With a carbon content of 55 percent (for bio-oil, the largest component), the amount of actual carbon entombed in asphalt, adjusted for reduced density from fossil sources, could be 270 million tons per year. Converting that to CO_2, we could be pulling more than 1.8 gigatons per year out of the carbon cycle just from road paving. Adding biochar to cement products alone is not large enough to reverse climate change, and the asphalt market is only half as large as the cement industry, but it all adds up.

CHAPTER 6

Fantastic Plastic

E ven though the unique properties of carbon make it a superior choice for plastics, up to now the high cost has been a challenge. What if that barrier could be breached by recycling and blending carbon from agricultural, municipal, and industrial wastes that might otherwise return to the atmosphere or ocean with older, recyclable plastics like polystyrene that are poisoning soils, waterways, and the ocean with a nondegradable toxin?

A composite is when two or more materials are combined to create a superior and unique material. The prefix *bio* means a composite blends natural fibers, including wood, leaves, and grasses, with a matrix (binder) made from either renewable or nonrenewable sources, such as lime, clay, plastics, or old tires.

Replacing Fossil Fillers

For most of biochar's ancient and recent history, it has been viewed predominantly as a "fillee" or carrier with countless pores to be filled with a variety of substances. Charging biochar's nooks and crannies with nutrients, either organic (e.g., urine, manure, compost) or synthetic, can transform it into a slow-release fertilizer. Saturating pore space with water can convert pyrolyzed carbon into something that can provide a low-cost, long-lasting irrigation pathway. Filling pore space with other substances, such as herbicides, can improve efficiency of the active ingredient while also reducing leaching.[1] Packing biochar pores with microbial inoculants has shown biochar can perform as well as or even better than other inoculum carriers such as peat moss or vermiculite.[2] Suffice it to say that carbon's role as a long-lasting pack mule for biological processes is well established.

The nonbiological processes are not nearly as well understood, but we are beginning to see carbon's potential as a filler for composites. Fillers have traditionally served to lower the cost of composites, but they can also improve various properties of plastics, paper, paints, and more. Fillers come in many forms; calcium carbonate, wood flour, and sawdust are among the most popular. Recent research has shown biochar can reduce costs while improving certain mechanical and electrical properties.

Take acoustic insulation, for example. Similar to thermal insulation, acoustic insulation materials are used in walls, ceilings, and floors to reduce noise. Typical materials are Rockwool, fiberglass, cotton, and cellulose. Rockwool sounds like a natural material but requires very high heat (1650°C/3000°F) to turn solid chalk and rock into something like cotton candy. The embodied energy is therefore very high: 4 billion calories per gram. While fiberglass, which is made of glass, sand, soda ash, limestone, borax, and binder coatings, is usually one of the cheaper options, it also requires a high-energy production and can have nasty health consequences, similar to asbestos. Cotton, mostly made from recycled jeans, is a great, albeit expensive insulator. Cellulose, which is cheaper and made from recycled paper, can only be safely composted at the end of its life if it is not doped with toxic flame retardants and insecticides. In terms of embodied energy, carbon compares favorably with most of these materials. Like cotton and cellulose, biochar can be composted at the end of its life. But biochar goes one step further: It can be easily molded into different shapes for optimal silencing or simply sprayed on like a plaster. Biochar as an acoustic insulation material could be a re*sounding* success.

Different polymer matrices in epoxy resins supply varying mechanical properties, chemical resistance, and thermal stability at low production cost. You will find them in glues, adhesives, surface coatings, and electrical insulators. Carbon nanotubes and graphene show great promise in boosting material performance of polymers, but are expensive materials. However, an important breakthrough is coming for commercialization of carbon-polymer composites in the form of cost reductions that can be achieved not only by cutting down the production costs, but also by reducing the cost of the filler.

An investigation of the mechanical and electrical (i.e., microwave) behavior of maple biochar as a filler found physical advantages over carbon nanotubes. The biochar particles fixed firmly in the matrix, were homogeneously dispersed, and hindered the accumulation of microcracks. During tensile load, biochar microparticles de-bonded and pulled out of the polymer epoxy resin matrix to help the matrix withstand higher loads while the large chunks formed a bridge between the matrix layers, creating an interlocking geometry with greater tensile strength. When tested for microwave percolation (filtration), the nanotubes were an efficient conductor, while the maple biochar was a strong insulator. Biochar's three-dimensional structure diminished particle interconnectivity, raising the percolation threshold to create a conductive path throughout the composite.[3]

Regardless of whether its role is as a filler or fillee, the added bonus of using biochar is that it can sequester carbon and, potentially, other undesirables (toxins absorbed from landfill leachates via phytoremediation, or heavy metals from mine and factory sites, for instance) for millennia. Carbon's ambidextrous character exponentially increases its potential as a carbon sequestration tool. What other material can boast this much bang for the buck?

Replacing Fossil Chemical Precursors

Australian author Tim Flannery, in *Atmosphere of Hope: Search for Solutions to the Climate Crisis*, recollects that before the oil industry developed, many of the products it generates today were extracted from wood. Lye, saltpeter, and potash, all derived from wood ash, were common in cleaning, food preserving, and explosives.[4]

The early wood chemists also knew charcoal could be made to yield various complex chemicals. Baked slowly in the absence of oxygen, its fumes were captured and condensed in a copper still and then decanted through a series of barrels and pipes. The result was a variety of fuels, solvents, explosives, dyes, antifreezes, preservatives, and early plastics such as Bakelite. By the 1930s, the wood chemistry industry was producing versions of most of the products that we get from fossil fuels today.

Flannery wrote, "Shortly after World War II, the industry collapsed due to competition from fossil fuels, which were becoming available in bulk and were transformable, on a massive scale, into fuels, fertilizers, and plastics."

The impacts on the forests were frightening and left lasting scars. By the late nineteenth century, an average wood chemistry plant in the northeastern

United States consumed 11,232 cords of wood per year. How many trees that translates to depends on size and species, but in the old-growth forests of New England at that time, a tree 20 inches in diameter and 70 feet tall would produce 1 cord, so each of those wood chemistry plants needed more than 11,000 trees per year. Forests were cut on a thirty-year rotation, which likely meant that on the second rotation foresters were settling for 10-inch-diameter trees 40 feet tall, or 70,200 trees per factory-year.[5]

Wood Chemistry

When the young alchemist Johann Glauber made pyroligneous acid in his apothecary shop in Amsterdam in the early seventeenth century, he may very likely have started with the tar, saturated it with chalk or slaked lime, and then heated and condensed the liquid to form an impure acetate of lime. Then, gently heating to destroy the oily matter without injuring the acid, he added the special ingredient that bears his name, Glauber's salt, generating a beautiful acetate of soda, in solution, which he drew off from the remaining sulfate of lime.

Not content to rest there, he heated it again, evaporated it to dryness, and redissolved and crystallized the acetate. These crystals he then placed in a retort with oil of vitriol and heated to vapor, which he captured and distilled into what he thought to be the active principle of vinegar, which he called *Acetum lignorum*, "wood vinegar."

With fossil fuels peaking, at least in some parts of the world, wood chemistry is coming back into vogue. Biochar production technology can be tailored to make more wood vinegar or less, depending on needs. The trade-off in materials produced (i.e., mass balance) includes tarry liquids, biochar, heat energy, and in some cases electricity. You can tweak the system to give you more of one or another of those, but it comes at the expense of the others. One thing it does not come at the expense of is forests. Wood vinegar can be produced from any substance capable of being carbonized.

This ability to shift the mass balance is actually one of the features that makes cascades antifragile. If, like sugar growers in Cuba, a refiner were to rely solely on the world market price for sugar, he would be destroyed if the price crashed. If the refinery could instead shift from producing sugar to producing power from biomass, or making alcohol, leaf proteins, pelleted fish food, biochar compost blends, or some combination of these products, the refinery could weather low sugar prices and ramp up sugar production again whenever prices rose.

Raw wood vinegar is as complex and variable as biochar. It has more than two hundred chemicals, such as acetic acid, formaldehyde, ethyl valerate, and methanol. Wood vinegar improves soil quality, eliminates pests, and controls plant growth, but can be toxic to fish and very toxic to plants if not sufficiently diluted. In certain cases, it may hold back plant growth even at small doses, but for many plants and soils it accelerates the growth of roots, stems, tubers, leaves, flowers, and fruit. A body of research shows that after an application of wood vinegar to an orchard, fruit trees increase the size and volume of fruit. Wood vinegar is relatively safe for food crops and for the birds, bats, and insects that pollinate them.

At the Asian Biochar Research Centre near Nanjing, China, a medium-scale, continuous-flow rotary kiln turns rice husks into biochar while generating 1.5 megawatts electric (MWe) of power for the local village. Wasting nothing, the plant gathers and bottles the flue gas and distills wood vinegar, which is sold as an organic pesticide and wood preservative. The kiln demonstrates complete flexibility in its process, able to transition from producing more biochar and less energy to more vinegar and less biochar, or any combination of the three.

In Glauber's time, the acid was used gastronomically as a substitute for wine or apple vinegar. A poor man's salt can be made by neutralizing the acid with a lye made from the ashes of the burnt wood. It was also used topically for treating wounds, ulcers, and other ailments. In fact, there are many health claims, but few are yet backed by solid academic research. Among the claims are that it supports digestion and combats acid reflux and heartburn; helps with colitis, Crohn's disease, and irritable bowel syndrome; supports liver health; cures diarrhea; arrests vomiting; balances the stomach's acidity; soothes peptic ulcers; fights tooth decay; has a positive effect on cholesterol and heart health; reduces swelling, redness, and itchy sensations from poison ivy; prevents infection from cuts, scrapes, and open wounds and by reducing bacterial accumulation prevents cellulitis; discourages ulcer-related infection such as circulatory issues in the arms and legs of elderly diabetics; stops ear infections; is a natural remedy for

The Benefits of Wood Vinegar

VerdiLife, a US-based wood vinegar producer, lists these qualities for its products:[6]

As a fertilizer:

- improves absorption through the roots
- stimulates plant and vegetable growth
- strengthens roots and leaves
- increases diversity of useful microbes
- increases soil microbial weight
- can be used as penetrant for better uptake: dilute 1:500
- increases crop resistance to adverse conditions
- improves tree health, producing darker green leaves for better photosynthesis, thicker and stronger stems, higher growth rates, making trees naturally more resistant to disease
- improves fruit quality and increases sugar content, flavor, color, firmness, and preservation in fruit
- can be used as flavor enhancement for agricultural end products: dilute 1:500 to 1:1000
- reduces excessive nitrogen levels, improving plant metabolism
- increases chlorophyll and elevates photosynthesis
- reduces artificial fertilizer

As a pesticide:

- repels pests, prevents plant infection from fungal, bacterial, and virus-like diseases
- inhibits virus and soil disease when mixed in high concentration
- prevents diseases caused by bacteria and fungi: dilute 1:200 and spray onto leaves

- repels parasitic nematodes: dilute 1:500 and apply to the base of plants
- can be used as corn preservative
- controls root rot: dilute 1:200 and spray onto leaves
- is harmless to bees

Other advantages:

- facilitates composting: dilute 1:50 and sprinkle on composting materials daily. This will help increase the biological activity of various beneficial microbes and can shorten compost times
- can be used for seed germination: dilute 1:200 and soak the seeds for twenty-four hours. More seeds will sprout
- reduces odor: dilute 1:50 to diminish the production of odor-causing ammonia in animal pens
- repels house flies: dilute 1:100 and apply to affected areas
- reduces incidence of chili pepper flowers aborting: dilute 1:300 and spray onto leaves

It is no small wonder with these benefits that the global wood vinegar market is expected to grow at a rate of 7.1 percent through 2025.[7]

gout by absorbing excess uric acid; and stops the swelling and infection of the prostate that leads to prostatitis.

What we do know for sure is that it can remove foul odors; works well as a pesticide, herbicide, and wood preservative; and in very moderate doses has been shown to stimulate plant growth. Whether it is actually good to ingest wood vinegar would depend on whether you are so desperate that you are willing to kill some of the good microbes that help your body perform vital functions and maintain your immune system to exterminate the bad microbes that are causing whatever might be your discomfort. To be safe, in our view, wood vinegar is best left to uses *outside* of the body.

Carbon Fiber

Until recently, all carbon fiber came from a chemical called acrylonitrile, made from petroleum, ammonia, and oxygen. The process for making acrylonitrile produced potentially explosive heat and made toxic wastes, including hydrogen cyanide gas. In 2017, a team of researchers at the National Renewable Energy Laboratory developed a process for producing acrylonitrile from corn stalks and wheat straw that doesn't produce heat and results in no toxic by-products.[8]

The MAI Carbon Cluster, an initiative from the German Federal Ministry of Education and Research, has been looking at high-volume production processes that could cut the cost of carbon fiber by as much as 90 percent and raise recycling rates to more than 80 percent. The effort, which has seen Audi and BMW working together despite initial reservations, now involves a total of 114 partners including Airbus, BASF, Eurocopter, SGL, and Voith.

During a 2017 workshop at Albert's Ecovillage Training Center in Tennessee, we made cascaded concrete with various biochar concentrations. We made composites by melting soy-foam packing peanuts and the kinds of Styrofoam clamshell containers they use for takeout in restaurants (and typically wind up in landfills, rivers, or the ocean). We made chardobe and biochar brick and compressed CINVA-ram brick. We made biochar grout for a tile bench. These exercises were only scratching the surface of the potential, but they showed what lies ahead. By melting extruded polystyrene foam packing peanuts and clamshell containers $(C_8H_8)n$ in an acetone bath $- (CH_3)_2CO -$ and adding powdered biochar (C) until it stiffened, we produced a light, structural, fracture-resistant char-tile that can be molded to any shape. It could be kitchen tiles, surfboards, iPhones, tennis rackets, boats, or biodomes.

The potential for these kinds of innovations is huge. In 2012, the global automotive industry produced about 63 million passenger vehicles and 21 million commercial vehicles. By 2020, production could grow to 100 million vehicles per year, with China accounting for up to 20 percent. The typical passenger vehicle curb weight ranges between 3,000 and 4,000 pounds (1,364–1,818 kg). The weight of sport utility and crossover utility vehicles (SUVs and CUVs) is usually an additional 500–1,000 pounds

(227–454 kg). Some quick arithmetic tells us that each year more than 150 million tons of new cars and trucks hit the roads around the world, embodying 120 million tons of steel and 10 million tons of aluminum. Currently, composites make up less than 1 percent by weight. Carbon fiber reinforced polymer (CFRP) makes up only about 9,000 tons, a minuscule 0.05 percent of the total global automotive materials.

Every 100-pound (45 kg) reduction in weight cuts fuel requirements by roughly 2–3 percent. And car designers have discovered that weight reduction in one area leads to weight reduction in other components and systems, resulting in a virtuous spiral of weight reduction. Composite bodies weigh 50–70 percent less (250 pounds/113 kg) than steel, which allows engineers to downsize chassis members; body panels and exterior accessories; structural and cosmetic interiors; suspension, drivetrain, exhaust, and engine bay pieces; brake systems; fuel systems; wheels; and other components. As weight becomes an increasing concern for designers wishing to optimize fuel mileage, as the impact of new carbon emissions regulation hits the steel and aluminum industries, and as the potential for automobiles to go from carbon producing to carbon removing is better understood, some big changes and opportunities lie directly ahead.

Old automakers who find themselves asleep at the wheel may find the marketplace is a cruel master. The penalty for not staying current increases with each design cycle, and design cycles are getting shorter, moving from about nine years to six or less. CFRP is an extremely strong and light plastic with carbon fibers woven in. When Elon Musk's SpaceX shuttle needed a way to reduce launch vehicle weight without compromising strength or other qualities, they turned to carbon fiber polymers. These are highly prized by many industries but at the moment they are very expensive to manufacture. High-end users like Ferrari or Jaguar can absorb the added costs and pass those along to their upscale clientele, but the same is not true for the mass market.

It is projected that by 2025, the auto industry, including race car teams and aftermarket accessory vendors, will consume about 25 percent of the global carbon fiber production capacity. Airlines could consume another 25 percent. Although carbon fiber polymer is a growth industry, there are drawbacks. Metals are readily repaired, reused, and recycled, and there is a huge global marketplace in all those areas. The same cannot yet be said of carbon fiber polymer. To avoid material wastes, landfill expenses, and exposure to fines in some regions, the industry is going to have to get a better grasp of carbon cascades.

Today, in 2018, the market for recycled carbon fiber (rCF) is 55 billion tons per year. Fifty billion of those tons could be filled with recycled scrap right now. That 5-billion-ton gap represents an immediate opportunity, but more important is the long term: designing recycling into the whole process. Lux Research cites present carbon fiber polymer capacity of 120,000 tons per year versus projected near-term demand of 225,000 tons per year, as rail cars, bridges, and buildings increase their carbon fiber content. A new carbon fiber polymer rail bogie (truck) frame is being made using 80 percent compression-molded recycled content and 20 percent virgin carbon fiber (vCF). These carbon rail cars reduce weight over their steel counterparts by 75 percent, cutting wheel-to-rail loads by 40 percent. The carbon fiber polymer industry must grow rapidly and rCF will play a major role in meeting demand.

There are market forces driving these changes much faster than the Paris Agreement or carbon credits can. Consider this hypothetical scenario, offered by an industry trade journal: Luxury automobile manufacturer X, which sells 100,000 vehicles annually in the North American market, can raise its average fuel economy from today's 29 miles per gallon (mpg) to 40 mpg by 2025, a 33 percent improvement. But it still fails to meet the US 55 mpg target. The fine assessed to the manufacturer is $55 per 1 mpg under the standard, multiplied by the manufacturer's total production for the US domestic market. In this scenario, manufacturer X would be fined approximately $82.5 million.

Similar incentives to meet such targets exist in Europe, but there the cost of failure is even more onerous. In the United Kingdom, failure to meet emissions standards results in a fine per gram of CO_2 per mile over the mpg limit. For high-emission vehicles like Jaguars, Land Rovers, and Aston Martins, this represents as much as an additional $20,000 added to the sticker price for each vehicle. If you are an automaker looking at these kinds of standards coming your way, you might want to research cascades that could get your product line out in front of these changes.

Retired or scrap carbon fiber for reuse in manufacturing is a first-stage cascade, easily accomplished by a combination of compression molding and thermoplastic films that provide shape and cohesion to the rCF content. A second stage could be separation of the carbon content in an exothermic process, burning or dissolving away the noncarbon portion and leaving behind cascade carbon that can be put to new uses. A third stage might be capturing the heat from that second stage and transforming it into process steam, electricity, or commercial heating and cooling. The refreshed carbon supplied by these processes offers scores of possibilities.

Carbon fiber polymers have been used in high-end automobile racing since Citroën won the 1971 Rally of Morocco with carbon fiber wheels. Low weight is essential for automobile racing, and carbon fiber is also ten times stronger than the steel it replaces. The substitution of lightweight carbon for heavier aluminum-lithium at the same strength gave SpaceX the ability to place a 300-ton reusable vehicle, potentially either an interplanetary spaceship or a cargo freighter, into low Earth orbit.

Racing-car manufacturers went on to develop omnidirectional carbon fiber weaves that apply strength in all directions, making the cars stronger than they had been when they were pure polymer. Building engineers, quick to adopt what they learned from Formula One, were soon wrapping carbon fiber polymers around steel-reinforced structures such as bridge or high-rise building columns. By enhancing the ductility of the section, they increased the resistance to collapse under hurricane, earthquake, or avalanche loading. In some countries prestressed concrete cylinder pipes (PCCP) account for the vast majority of water transmission mains. Due to their large diameters, failures of these pipes are usually catastrophic and affect large populations. Now carbon polymers are being retrofitted as PCCP liners that take strain off the host pipe.

As recently as seven years ago, automaker BMW was using water cutting for parts, but today, in partnership with Airbus Helicopters and others, the carmaker has moved to carbon-cutting tools coated with ground diamond that can double feeding speeds. The carbon tools have a geometrically defined cutting edge and are sharpened by a plasma process. For BMW and Airbus, production costs are reduced 90 percent. Bicycle frames of carbon polymer give the same strength as steel, aluminum, or titanium for much less weight and can be tuned to address different riding styles. Carbon fiber cellos, violas, violins, acoustic guitars, and ukuleles are selected by discerning musicians for the quality and fidelity of their sound. Other commercial products with carbon fiber already available include:

- bagpipe chanters
- billiard cues
- carbon fiber posts in restoring root canal–treated teeth
- carbon-woven fabrics
- drones
- drum shells
- fishing rods
- guitar picks and pick guards
- helicopter rotor blades
- high-reach poles for window cleaning
- laptop shells
- loudspeakers
- passenger train cars and furnishings
- suitcases and briefcases
- tent poles
- thermoplastic films for moisture and corrosion barriers
- tripod legs
- turntables
- violin bows
- walking sticks

Combining biochar at rates of 5, 15, 25, and 40 percent by weight with wood and plastic to make alternative composites to traditional wood-polypropylene binders, scientists found:

- All biochar rates increased flexural strength by 20 percent or more.
- Tensile strength was highest with 5 percent biochar.
- Tensile elasticity was highest with 25 percent and 40 percent biochar.
- Water absorption and swell decreased.
- Biochar additions showed improved thermal properties.

Wood–plastic composites (WPCs) have annual growth rates of 22 percent in North America and 51 percent in Europe. Often polyethylene, polypropylene, and polyvinyl chloride use wood flour or fiber as fillers, and more recently, resin-impregnated paper waste from particleboard and fiberboard manufacture. The advantages of using bio-based components in these plastics is that wood and paper are nonabrasive, low in cost, widely available, low density and weight, flexible, and recyclable.

Impregnated paper waste is a major challenge for recycling due to the large amounts produced, potential toxicity, and low biodegradability. A medium-sized paper impregnating factory will produce 400 tons per year. One option is oriented strand board, but that just kicks some of those problems down the road. The strand board can still off-gas toxic chemicals. A better option could be pyrolysis, followed by incorporating the pyrolysates into a long-storage option.

Decking for outdoor applications represents the largest market for WPCs. In Europe, that market, outside automobiles, is 120,000 tons, with more than half going to decking. Manufacturers are currently shifting product lines to include siding, roofing, windows, doorframes, and outdoor furniture. Some are already incorporating nanoscale reinforcing fillers like nanoclay and carbon nanotube into the composite material.

An extrusion technology called waxy technology recycles and transforms more than twelve different types of postconsumer plastics and packaging materials into long-lasting, termite-resistant plastic lumbers, potentially sparing many forests from the ax. An ideal product for building, construction, and furniture making, extruded lumber costs 32 percent less than pressure-treated timber, avoids arsenic and other ecotoxins, and lasts more than forty years without replacement even in sunny, windswept, and coastal areas, or in underwater applications. Applying cascade carbon thinking to this scenario could supply both process heat and a low-cost, high-value filler material, and sequester ever more carbon.

Any carbon that does not go back to the atmosphere or to the oceans can take a break from the carbon cycle. It doesn't have to burn to become carbon dioxide. It doesn't have to digest or decay to become methane. It doesn't have to kill coral reefs or warm the Earth. It can just chill. It can be a building or a bicycle, it doesn't matter. Just chill a few centuries while humans recalibrate. Carbon, arranged into chains and rings by photosynthesizing plants, then rearranged to weave into fabrics, fibers, and filaments, will soon surround us in our buildings, modes of transportation, and much, much more.

CHAPTER 7

Paper Chase

Typically about a third of all waste generated in an industrial country is paper and cardboard. With landfills reaching capacity and methane emissions a growing concern, many countries have begun to restrict or reduce the use of these materials. One emerging solution is to pyrolyze the paper or cardboard and then upcycle it back into secondary uses. Biochar paper and biochar polymers are two products that can find immediate application in several industries. They simultaneously store carbon and improve the products they are part of.

Pack your fruits and vegetables in a biochar box and you may double their shelf life. Then compost the boxes, along with the leftovers, and make gardens in your backyard. Or slip a thin biochar-paper shield under the light switch and electric socket covers in every room. You are protecting your family from radiation while upcycling trash.

Allow us to introduce you to chardboard, a blend of biochar and paper pulp Kathleen created with the assistance of a paper designer from the Genesee Center for the Arts in Rochester, New York.

Paper and packaging materials generally contain a mix of fiber and filler material. Fibrous material acts as the binding agent while fillers have traditionally been added to lower production costs. Choices of fillers have varying particle size and shape, surface area, and color to optimize bulk, printability, stability, drying rates, and other qualities. The end use for the paper product determines which properties are most important. Why not cascade carbon fillers?

Zion Market Research predicts that the global green packaging industry will grow to more than $242 billion by 2021.[1] Driven by public demand for green products, manufacturers and retailers have been exploring packaging systems that use renewably sourced materials and anticipate waste recovery. Biochar paper or cardboard meets those criteria, and can be produced from any number of woody wastes, including wastes from the pulp and paper plants themselves.

Paper mill effluent is a contaminant to surface water, so most of it (72 percent) is dehydrated and landfilled. Biochar could filter the effluent and then be added to paper products that do not harm people or the environment. Call that cool paper.

Biochar has other functional properties for packaging. The porous quality of pyrolyzed carbon gives thermal insulating properties to cardboard. By absorbing ethylene, chardboard can slow ripening in produce. It can keep fabrics clean and fresh. It will protect electronics from electromagnetic fluxes. It absorbs odors and condensation. Carbon paper fillers can replace many of the synthetic products in packaging and will continue to benefit the environment long after the package is no longer in use even if, or perhaps especially if, it ends up in the landfill.[2]

Many municipalities are striving to divert food wastes from landfills into composting centers where nutrients can be recovered and sold, but they are challenged by the odors typical of decaying food waste. If disposed at curbside in easy to collect, bio-upgradable containers, chardboard could reduce odor from garbage on its way to becoming compost. Once in the compost pile, it speeds the compost process, reduces nitrogen losses, and improves the C:N ratio needed for effective composting.

Some fruits and flowers are highly susceptible to premature ripening due to ethylene, a natural plant hormone that promotes cell degradation. Carbon effectively sorbs ethylene – for this reason carbon molecular sieve membranes outperform polymeric membrane performance in separating ethylene gases.[3] Biochar added to packaging for fruits such as bananas or kiwis could help extend their shelf life. Its porous nature allows it to absorb moisture and vapors and keep products fresher longer. If the biochar is made to optimize hydrophobicity, it could potentially resist fungi and molds as well.

Electrostatic discharge can damage electronics components during shipment. Biochar produced at higher temperatures is conductive and antistatic,[4] so chardboard would dissipate charges before they build up, as well as protect electronics from mechanical damage by cushioning them. That reduces the necessity for multiple packaging elements.

Customized paper blended with charcoal is not a new technology. Japanese companies have long been using a charcoal-based tissue material called *umezumi* made from carbonized plum pits capable of deodorizing and absorbing volatile compounds and moisture. Experience in Japan with biochar tissue paper shows biochar's ability to adsorb odors after being blended with paper pulp. This kind of paper is used for shoe soles, drawer or refrigerator liners, and mattress pads.

Electrosmog Defense

From the moment we awaken to the time we go to bed, and all through the night, most of us in the industrial world are bathed in an electronic soup. Even when our minds shut down for sleeping, our bodies are being bombarded with electromagnetic radiation — radio and television signals, radar from passing airplanes and satellites, microwave ovens, cellular phone networks, nearby Wi-Fi hubs, Bluetooth pairings, electric currents in walls and computers, sensors in automatic doors and lights, smart refrigerators and showers. Pass through an airport or government building and you may be X-rayed, magnetically scanned, and wanded. Stand in line at a checkout counter and the radiation from the scanners, cash registers, credit card terminals, and RF tags is going through you all the time. As the internet of things exponentially emerges, we are being irradiated by wristwatches, wallets, car seats, running shoes, and clothing.

It is too soon to say what all this is doing to us biologically, but for some, getting away from it or just turning it off has become a regular ritual. If you live on Earth in the twenty-first century, you cannot elude it all, even if you are part of an undiscovered tribe in the upper Amazon or Papua New Guinea. A Himalayan yeti or Yukon sasquatch might not understand Amazon Echo, Google Home, or Apple HomePod, but they are nonetheless bathed in electromagnetic frequencies.

In 2017 we hosted a weeklong workshop called Biochar from the Ground Up. We'd been researching biochar and cement for a few years, had blended char with other synthetic and organic materials, and Kathleen had brought with her some homemade paper samples impregnated with fine powder biochar. It occurred to us that the biochar in the paper might block the electromagnetic field from the wall current in the classroom.

Albert took a piece of the paper over to a wall socket and set up an experiment. First, he took an electromagnetic field strength reading about 10 centimeters (4 inches) from the socket. The meter pegged at over 100 milligauss. Then, passing the biochar-laden paper between the socket and the meter, we watched the gauge immediately bounce to zero. As he withdrew the paper, the meter glided back to the top of its range. Just to be certain it was the biochar and not the paper, he repeated the experiment with a plain sheet of unimpregnated paper. It had no effect. The meter remained at the top of its range. We posted a video of the experiment to Instagram.[5]

Besides demonstrating what we already knew — that porous carbon blocks electromagnetism (and ultraviolet, infrared, and most forms of nonionizing radiation) — we showed it takes only a paper-thin layer to be effective. It has been shown in paints and plasters that adding biochar has this effect, reducing radiation from electric power lines, transformers, and Wi-Fi modems. It also brings up an interesting idea for yet another carbon cascaded product: impregnated papers.

With the predominance of the internet and electronic technologies, traditional papermakers are feeling the crunch. Those that want to stay in business are searching for new outlets: luxury or high-end products, digital printing, and "the papers of the future." In 2010, Les Papeteries de Vizille in France started producing scented papers for Hermès and Yves Saint-Laurent catalogs. Looking ahead, they are working with laboratories of the CEA (Commissariat à l'Energie Atomique) in the nanotechnology and nanomaterial fields to develop smart papers that are capable of working in high-radiation environments.[6] Can radiation-blocking insulators for electric switch covers, junction box covers, and personal computers be far behind?

Given the ubiquity of electronic smog and the need to prevent damage — to vulnerable humans and other living things, equipment in sensitive installations like laboratories, hospitals, radiation treatment centers, nanotechnology manufacturing facilities, and space missions — potential applications for this new carbon product are boundless.

An initial investigation by Kathleen used biochar derived from coffee chaff, the skin of the coffee bean removed before roasting. Coffee chaff char is relatively uniform in size and composition and qualifies as one of many sources of "unloved biomass" in that it is often shipped off to landfills at great expense to coffee producers. The first tests used a range of filler content in paper varying from 0 percent to 30 percent.[7] Paper pulp was the only other material used. The chardboard had several interesting properties, including low thermal conductivity. Reduced heat transmission is helpful for hot or cold food delivery such as Chinese noodles, pizza, and sushi.

Many biomass boilers may be subject to emissions standards that require best available control technology for pollutants. More than half of all paper mills use biomass boilers for some portion of their energy needs. Consequently, many aging boilers are now slated for replacement. These are good candidates for conversion to biomass boilers or combined heat and biochar.

Paper Cascades

Coffee chaff biochar can be turned into coffee drip percolation sleeves; carbonized citrus peels can be converted into chardboard frozen juice containers; paper sludge, sawdust, mill wastes, and bark can be charred and put back to the paper manufacturing process, while generating heat and power. The number of possible uses and benefits for chardboard seems almost limitless, even without accounting for carbon sequestration when the products eventually return to the soil. Here are some possibilities:

FOOD AND BEVERAGE

Bottle labels
Coasters
Coffee cup sleeves
Coffee filters
Coffee packaging
Cup holders
Disposable plates
Egg cartons
Food trays
Fruit packaging
Napkins
Placemats
Sandwich wrappers
Take-away containers
Tea bags

FARM AND GARDEN

Leaf litter bags
Plant wraps
Seed blockers
Seed tapes
Seedling starter pots
Sheet mulch
Weed barrier cloth

HOME AND OFFICE

Air filters
Book covers
Bookmarks
Ceiling tiles
Computer covers
Drop cloths

Furniture
Hangers
Kitchen waste compost bags
Picture matting
Toilet covers
Wall coverings

MEDICAL

Band-Aids
Biodegradable bedpans
Biodegradable disposal bags
Biodegradable kidney dishes
Biodegradable urinals
Gurney liners
Patient gowns and slippers
Skin wraps
Surgical attire (masks and caps)
Underpads and exam table paper

SHIPPING SUPPLIES

Bags
Boxes
Cubicle dividers
Edge protectors
Envelopes
Fiber drums
File dividers
Folders

Inboxes and outboxes
Notebook covers
Shipping tubes

PERSONAL USE

Biodegradable urns
Diapers
Gift wrap
Paper towels
Sanitary napkins
Toilet liners
Toilet paper

PETS

Bird cage liners
Fish tank filters
Pet caskets
Poop bags
Kitty litter liners
Wee wee pads

MISCELLANEOUS

Car insulation
Caskets
Casket liners
Gaskets
Newsprint
Wall insulation

As with other areas of biochar technology, cost is often the biggest hurdle to adoption, but carbon offers multiple cobenefits, particularly in closed-loop production scenarios. Industries that generate significant amounts of biomass waste and that have a need for packaging material — pulp and paper, coffee, and fruit, for instance — may be especially well positioned to take advantage. Imagine turning the waste from an agricultural crop into packaging for the same crop! But now that we mention it, this is not that new. Mother Nature was already doing that.

CHAPTER 8

Filtration Nation

Declining water supply and quality may be a more immediate threat than climate change, although the two are not easily separated. We should be preparing to deploy biochar for filtration and as a desiccation medium widely after major weather events to prevent them from escalating into public health emergencies. Better yet, we should be preparing to make the biochar from downed trees while supplying heat and electricity to areas that have lost power.

In a natural disaster, one of the greatest threats to drinking water is flooding. Climate change has already begun to change both the frequency and the intensity of precipitation events with hundred-year, or even thousand-year, floods occurring several times each decade. Floods are both the most common and the most costly natural disaster. They lead to erosion, landslides, the overwhelming of sewers and water treatment plants, and destruction of vital barriers that protect human health from exposure to contaminants. This is why outbreaks of deadly diseases often follow when floodwaters recede.

As an increasingly unstable climate brings warmer temperatures and more and bigger floods, there is a risk of infectious pathogens spreading to millions more people around the world. Pathogens come in many forms — bacteria, fungi, viruses, prions, and protozoa — with varying degrees of toxicity. Even in an industrialized nation like the United States, the national Centers for Disease Control and Prevention (CDC) estimates some 48 million people get sick from food they eat, with 128,000 hospitalizations and 3,000 deaths each year. The annual health care costs for E. coli in the United States alone are estimated at $480 million.

Biochar barricades can be used at treatment facilities and along runoff areas to reduce the spread of dangerous pathogens in high-water events.

Temporary floodwalls made of carbon-based geotextiles and filled with biochar can be rapidly deployed to hold back floodwaters and later recovered for reuse or upcycling. Alternatively, biochar can be spread on flood-damaged fields to help absorb toxins. Biochar has the added benefit of adsorbing heavy metals and nitrates and then remaining behind to rebuild the soil's biological wealth once the waters recede.

Research from McGill University shows biochar-amended soils can reduce leaching of fecal coliforms into groundwater or surrounding local water bodies.[1] The USDA's Food Safety and Intervention Technologies Research unit found that various different types of biochars can inactivate *E. coli* in certain types of soils amended with 10 percent biochar by weight. Researchers at Stanford, USDA, and Berkeley discovered that adding biochar to biofilters can significantly decrease the amount of *E. coli* carried in stormwater.[2] Higher-temperature chars and wood-based chars appear to be the most effective. Chars with more hydrophobicity (ability to repel water) also appear more effective. Particle size and application rate can be important.

In the case of a field or garden contaminated by floodwaters, heating and drying can be accelerated by spreading finely pulverized biochar over the surface and allowing sunlight to work its magic. A temperature above 60°C (140°F) will result in near instant kill for most pathogens excreted in feces. Temperatures in the range of 50–60°C (122–140°F) result in no growth for bacteria and death, usually within thirty minutes. Decreasing moisture and increasing temperature can work together to produce a faster die-off.

Using charcoal to remediate toxins is at least as old as using it to improve soil fertility. It is still a common practice in some parts of the world. Organizations like Aqueous Solutions in the United States and Thailand are teaching people around the world not only how to produce their own carbon media, but how to set up filtration systems capable of capturing many common contaminants. They have shown high-quality biochar for filtration can be easily made using locally available biomass and low-cost kilns.

For a planet covered by more than 70 percent water, it is shocking to witness the toll water scarcity takes. Fresh water is only 3 percent of all water, and that includes the part melting off Greenland and the poles right now. Pumping fossil water from the pores of deep rock while simultaneously paving the landmass needed to recharge aquifers leads to sinking cities and thirsty people. Sending toxic wastes from farms, industries, and sewers into

Water-Stressed Cities

In the weeks and months after Hurricane Irma, Puerto Rico struggled to restore one of its most precious commodities — fresh water. Stores became empty soon after the storm. With Puerto Rico's bottling and water treatment plants wrecked or without power, it was unable to deliver tap water and had to rely on deliveries from the outside that could not possibly keep up with what was needed.

On the other side of the world, Jakarta is sinking faster than any other major city. The problem is less about climate change and the rising Java Sea and more about unchecked development — a near-total lack of planning; inadequate sewers constructed as an afterthought; and a fragile supply of imported water. When the Dutch arrived in the seventeenth century, thirteen rivers fed the city. Today, faucet by faucet, people are draining aquifers dry. When it rains, the rivers overflow. Mangroves that used to absorb and cleanse have been erased and replaced by shantytowns and apartment towers.[3] It's the tragedy of the commons, on steroids. About 40 percent of Jakarta has sunk below sea level.

Mexico City is sinking even without nearby seas or hurricanes to blame.[4] The city's need for water long ago grew past the natural availability of water, even for a city built on an ancient lake. Now the lake sediment is part of the problem. It cannot support the weight of such a large city. Pumping the underground aquifer to slake the thirst of 21.5 million people has made the lakebed subside. And not just a little. In a plaza near the Monument to the Revolution a water pipe now arches 26 feet (8 meters) overhead. When it was placed into the earth in 1934, it was below the street and below the shallow aquifer. In the last eight decades, while the plaza has slowly dropped, the pipe has held its original position. People wander into the plaza and stare up at this strange, out-of-place pipe.

Here is one of those statistics that leave you either uncomprehending or stupefied: By 2025, *half* of the world's population will be living in places that are as water-stressed as Puerto Rico, Jakarta, or Mexico City.[5]

Threats to safe and available water are not limited to sinking cities or drought-prone regions. Eight hundred forty million people lack even a basic drinking-water service, including 159 million people who are dependent on surface water. Two in seven people in the world use a drinking-water source contaminated with feces. In low-income and middle-income countries, 35 percent of hospitals and clinics lack water and soap for hand washing.

Biochar can help prevent contamination and can help decontaminate polluted water bodies. It can help lessen flood risk by creating urban sinks for water and recharge aquifers. As attractive as biochar is as a climate mitigation tool, its ability to ensure access to clean water, goal number six of the UN Sustainable Development Goals, may be even more essential.

whatever bodies of water will wash them away has filled lakes and rivers with liquid that sometimes bears little resemblance to what we call water.

Land Reclamation

Humans have been extracting various minerals, metals, and other materials from the Earth since the dawn of history. As we've "advanced," our methods of mining have gotten progressively more destructive; abandoned mines often resemble an open sore, the mountain spine broken, colored streams bleeding downhill when it rains, and outgassing methane for decades.

We remove mountains for crushed rock, cement, sand, coal, copper, gold, and iron. Coal alone has led to more than 6 million acres in the United States being turned into a moonscape. If all mining activities were concentrated in one place, it would fill an area about the size of Belgium.[6] Although many nations have laws requiring companies to set aside money to restore and revegetate denuded mine lands, regulatory enforcement is often lax and mining companies find loopholes. In the United States alone, the unpaid cleanup bill is $50 billion and rising.[7]

Chris Peltz, a researcher and practitioner with Research Services LLC in Colorado focusing on land reclamation, was initially skeptical of biochar's ability to help restore landscapes damaged by mining, drilling, and other human-caused destruction to land. Eight years of trials, from small (0.5 acre, or 2,000 square meters) to substantial (many acres), have changed his mind. Peltz believes that arid sites or soils subjected repeatedly and over the long term to all manner of "cides" (pesticides, fungicides, herbicides), or places where other remediation efforts have failed, are all good candidates for carbon to convert adversity to advantage.

Many reclamation efforts are challenged by costs. To address this hurdle, Peltz recommends a carbon cascade using old pallets to create heat and biochar, filling some type of filtration container such as a bag, sock, or pillow, and placing it in a settling pond or basin to treat metals or for acid mine drainage. Next, the iron-rich biochar could be used on livestock effluent; the metals will actually enhance adsorption of certain nutrients such as phosphorus and nitrogen. Then, saturated with nutrients, it can be used to revegetate mine lands, plant urban trees, or revegetate highway margins for stormwater control. And at that point, the price of biochar will be more in line with tight reclamation budgets.

Contaminated land should not be used for growing crops, but in some places farmers do it anyway because they have little choice. They keep quiet, knowing that they won't be able to sell their crops if consumers know they're grown on contaminated land. In China, the government found 16 percent of all land or 19 percent of farmland (250,000 square kilometers — 96,526 square miles, roughly the size of Michigan or the United Kingdom) was contaminated by pollutants or metals such as lead, cadmium, and arsenic. More than a quarter of all rice tested had excessive amounts of lead. Pharmaceuticals, pesticides, and personal care products pollute soils, groundwater, rivers, and coastal estuaries around the world. It is a massive cleanup job, but also a massive carbon sequestration opportunity.

Metals and other contaminants in soils are only toxic if they are mobile, meaning they can either move within the soil profile and possibly contaminate groundwater, or they can be taken up by plants. That can be either a blessing or a curse. Phytoremediation is a technique used to decontaminate soils, water, and air that takes advantage of certain plants' ability to absorb toxins. These plants are later harvested, leaving soils less laden with contaminants. Phytoremediation can be a relatively low-cost technique but does not necessarily prevent air and water contamination, depending on what becomes of the harvested plants. Pyrolyzing the harvest can stabilize the contaminants, entombing them within the pyrolysate's pores. Biochar can both help plants grow in adverse conditions and later dispose of the soil contaminants absorbed into the trees, which once carbonized can go into hardscape structures as aggregates or carbon fiber.

This idea is being put to the test in China. The Chinese government is trying not only to reduce toxins in soils, but also to reduce air pollution caused by in-field burning of crop residues, making biochar a natural fit on both sides of the pollution problem. Through surface adsorption and precipitation, different types of biochar or biochar composites can mitigate plant uptake of cadmium, lead, copper, zinc, arsenic, and polyaromatic hydrocarbons while also boosting yields in rice. Add the ability to sequester massive amounts of carbon and generate renewable energy, and you've got a winning combination.

Contaminated soil around landfills amended with biochar and green waste compost will, within about sixty days, chelate (bond to) copper and arsenic and sequester zinc and cadmium.[8] Adding biochar to an arsenic-contaminated soil favors the growth of fungal mycelia that, creating an insoluble wax of arsenic compounds, reduces the uptake to plants.[9] Biochar can bring a tenfold decrease of cadmium to groundwater. It also has shown greater than 50 percent decrease of the heavier, more toxicologically relevant polyaromatic hydrocarbons.[10]

One of the things field trials also show is that highly alkaline and water-soluble forms of carbon such as some types of compost, sawdust, or wood chip mulches can undesirably mobilize certain elements. Pyrolysis changes this. If biochar is blended with compost, revegetation and recovery happen rapidly, usually with far less toxic leachates.[11]

Heavy Metals and Radionuclides

Apart from nuclear waste products, the four metals most toxic to humans and other biota are cadmium (Cd), mercury (Hg), lead (Pb), and arsenic

French Drains

While roaming the home improvement stores, Kathleen noticed the fabric sock made to keep the perforated pipes used in foundation work from filling with silt over time, which would defeat their purpose. French drains and curtain drains are normally installed whenever there is a grade difference that puts dirt up against the foundations of a building. Perforated pipe, laid below the level of the floor slab, is surrounded and covered with porous media such as sand or gravel, gathers groundwater, and channels it to daylight and a swale or somewhere downslope.

There in the landscaping aisle, Kathleen noticed "these weird things that looked like giant worms," which turned out to be Styrofoam packing peanuts inside the drain sleeve surrounding perforated pipe. "Perfect" for French drains, the advertising claimed, which might also mean how perfect it is to assemble $15 worth of ingredients and make $49 retail. Why use extruded plastic peanuts when raw biochar, in chunky sizes, could provide similar air cavities while also filtering the groundwater of pesticide residues, nitrates, and other undesirables?

Then we started thinking, "Let's ditch the drain sleeve." Just surround the slotted pipe with biochar and topsoil. Then we got to wondering how much you really need PVC pipe if you have a foot or more of pretty clumpy biochar. And while we are at it, perhaps we can create culverts out of porous carbon instead of cement or plastic. Imagine how many toxins might be prevented from getting into local water bodies. Imagine how much carbon you could sequester if you did this for all building construction sites and hardscape features.

(As, actually a metalloid). Excessive amounts, in readily bioavailable forms, are spread around the global commons in our air, oceans, lakes, rivers, soils, and sediments. They are sprayed on plants, belched from smokestacks, and left behind at mines and factories. No one is immune because our air, water, and food travels around the globe, exposing every living thing on the planet. Carbon may be our best hope for containing this chemical catastrophe.

Many, though not all, types of carbonized organic matter can stabilize heavy metals in soil and remove them from aqueous solutions. Carbon in the form of living plants, animals, insects, and microbes performs remediation above- and belowground simply as part of what it does for a living.

Coal-fired power plants are the single largest source of mercury contamination. Nuclear plants, and their supporting fuel cycle, are the single largest source of radioactive contamination. While eliminating coal and nuclear power plants entirely would be the best way to prevent them from poisoning the environment, capturing emissions at the source supplies a temporary patch to reduce the damage. Most remedial treatments are costly and so, not surprisingly, the worst sites all too often go untreated once they are retired.

Magnetized biochars have proven more effective than certain activated carbons in removing mercury from power plant exhaust. Cleaning up mercury that has already landed in water and on land can be complex and often yields short-term or less-than-satisfactory results. Recently, however, high-temperature biochar blended with sulfur was found to remove more than 90 percent of mercury in aqueous solutions[12] and can reduce transport of mercury in contaminated sediments by more than 80 percent.[13]

Cadmium also comes from fossil power plants but more often gets into the food chain via chemical fertilizers. It can lead to kidney failure and other serious health problems. Biochars can reduce cadmium mobility and bioavailability,[14] preventing it from entering the food chain while simultaneously improving crop yield. New regulations limiting the amount of cadmium in chocolates that will be imposed in Europe in 2019 have put many cacao farmers on notice to find solutions to lower cadmium levels or lose access to an enormous market. Dr. Brenton Ladd, a transplanted Australian teaching environmental engineering in Peru, is showing rural farmers how to make and use magnetized biochar to immobilize metals and maintain yields.

Another crop experiencing the downside of fertilizer is tobacco, a demanding plant for soils. For nearly a hundred years, super-triple-phosphate fertilizers have been applied to keep up annual production. In the early 1970s, Edward A. Martell, a former nuclear weapons test manager for the air force who had relocated to the National Oceanographic and Atmospheric Administration in Boulder, Colorado, discovered that because of where they were mined, super-triple-phosphate fertilizers contained high levels of natural thorium.[15] When added to soils, typically side-dressed onto rows of tobacco, the fertilizer's thorium would break down by radioactive decay into radium and then quickly disintegrate into a trio of toxins — radioactive lead, bismuth, and polonium — that would drift out of the soil as gases but because of their weight would remain close to the ground and be deposited on the sticky underside of the tobacco leaf.

What happens in the bodies of smokers is easy to imagine: Ignited tobacco smoke passes down the airway, first burning away the epithelial hairs on the bronchial passages so that tars could thereafter not be easily trapped or ejected, and then lodging in the lungs of smokers as radioactive source material, continuously irradiating cells during their most sensitive moments of mitosis when genes were opened and most vulnerable to mutation. Result: lung cancer. Of course, cigarette smoke is not the only cause of lung cancer, nor are the radon offspring the only elite ninja assassins hiding in a smoker's lungs. We are all exposed, all the time, to mutagens and carcinogens. The toxic soup we are bathed in gets thicker by the year.

Lead has featured in recent headlines because drinking-water sources for entire communities were built in past centuries of pipes and fittings made of lead alloy. Thousands of communities are being contaminated from out-of-date infrastructure. As we've seen in the tests conducted at wastewater treatment plants, biochar as pre- or post-treatments from kitchen sinks to highway runoff has been shown to be effective at removing lead.

Recalcitrant carbon in the form of biochar and labile carbon such as green waste compost make a good team for decontamination. The two have opposing specific metal attraction, so in combination they catch a broader spectrum of pollutants and still dramatically increase plant growth.[16]

A group of soil scientists led by Dr. Genxing Pan at Nanjing Agricultural University used biochar amendments to reduce N_2O loss from fields of grains and rice. They also studied the effect on reducing heavy metal uptake and accumulation in crops grown in polluted fields. In southern China, rice grown in regions with high soil cadmium is becoming a public health concern. Pan succeeded in developing an effective amendment to ameliorate cadmium uptake in rice and went on to demonstrate its ability to repair sodic (high salt) soils in the wheat-growing valley of Yellow River, in Henan Province. In wheat fields with salt content above 0.5 percent and pH over 10 (very alkaline), Pan employed an amendment of biochar charged with wood vinegar and obtained a normal wheat yield (6.7 tons per hectare, or 2.7 per acre)

Countering Air Pollution

Breathing clean air is not always an option. This has given rise to the disposable face mask trend in some cities. The sources of bad air may vary. Some can be traced to natural sources like mildew, wildfires, and volcanoes, for instance, but most air pollution today stems from human activity. Urban environments suffer from car pollution while rural areas are afflicted with crop burning, pesticide drift, and ammonia from animals and fertilizers. Wind and weather can turn local air-quality problems into larger regional or sometimes global problems.

Biochar cannot counteract all these types of air pollution, but it can help with some. Farmers commonly set fire to rice or other crop residues after harvest, the cheapest way to clear fields. While the governments in China and India have moved to restrict in-situ burning, enforcement of such regulations is sporadic at best, so farmers often only risk the small possibility of a fine. Portable smokeless kilns could easily convert much of India's 32 million tons of leftover straw into biochar if kilns were more easily available.

Nanjing Agricultural University has been working closely with Chinese companies to collect, pelletize, and carbonize straw using large-scale, centralized pyrolysis units. Given the small plots typical of Chinese agriculture, collection of biomass is no easy feat. Beijing Sanju Company, seeing the opportunity, has developed a viable business model. It rents land from small farmers to consolidate the supply of straw. As of 2017 Beijing Sanju was operating three production facilities, each capable of producing 30,000 tons per year of biochar. After more than five hundred biochar trials, they have demonstrated to rice farmers that yields are better, shoots come up quicker, and plantings have more shoots. They have also shown reduced metal uptake by crops.

while neutralizing soil contaminants. Before the treatment, the region's soil produced yields of less than one-tenth of Dr. Pan's and efforts to restore organic constituents in the soil had failed to revitalize the fields.[17]

Table 8.1. Carbon Math: Rice and Wheat Straw in China

Straw produced in China	7.0 Gt/yr
Amount burned in the field [25%]	1.75 Gt/yr
Covert 25% to biochar	0.44 Gt/yr
With 60% C content	0.26 GtC/yr
CO_2-e drawdown potential	(0.96 $GtCO_2$-e/yr)
CO_2-e avoided	0.5 $GtCO_2$-e/yr

Note: Potentials for atmospheric removal are in parentheses. The degree of substitution is based upon the physical properties of the substance being displaced and the authors' estimate of market penetration potential

In Shanxi Province, a group of farmers developed a mobile pyrolysis system that can be driven by a tractor to the fields to convert maize, wheat, or rice straw into biochar. The mobile kiln produces 120–150 kilograms (264–330 pounds) of biochar per hour using the crop straw produced in roughly 1 mu (7,200 square feet, or 669 square meters). Because burning straw in the fields accounts for carbon emissions equal to 15 percent of that from China's energy production, this cascade solution is not insignificant. Pan calculates that if adopted widely, carbonizing crop residues could reduce China's CO_2 footprint by 200 megatons annually while also enabling the production of healthier food.

Respectfully, we think it could be much more than that. China's agricultural soils are estimated to have less than a third the soil carbon of United States and European agricultural soils, leaving a huge capacity for soil carbon accumulation. The combination of benefits — neutralizing toxins while also sequestering carbon — lends hope to reversing the carbon cycle in the world's most populous nation and largest carbon emitter. We ran the carbon math for straw residues from China and found that if a quarter of the residues could be charred, the drawdown potential is enormous — almost a billion tons per year. Carbonizing in lieu of burning would wipe out nearly a tenth of China's carbon footprint. Add the avoided emissions from crop burning and that number climbs to almost 15 percent.

CHAPTER 9

Cattle Carbs and Leafy Greenbacks

I n the second century, Cato the Elder wrote in *On Agriculture:*

> If you have reason to fear sickness, give the oxen before they
> get sick the following remedy: 3 grains of salt, 3 laurel leaves,
> [. . .], 3 pieces of charcoal, and 3 pints of wine.

We can't speak to the benefits of serving wine to your cattle, but livestock farmers have many options for using biochar throughout the production cycle, and the earlier it starts, the more cascading benefits farmers will reap.

Until the beginning of the twentieth century, supplementing with charcoal was common practice to increase animal performance and health. During the last century, veterinarians continued to investigate using activated carbon as a treatment for poisoning, bacterial infections, and viral diseases. That all ended when transnational pharmaceutical companies branched out from human medicine to animal medicine in the second half of the twentieth century. Livestock now accounts for 80 percent of the antibiotics market. Healthy animals are fed pharmaceuticals on a daily basis to prevent illness. Antibiotics can also help them gain weight more efficiently. Biochar can do both, and feeding biochar to cows and other livestock will not contribute to either antibiotic resistance or the growth of superbugs.

Around 2010, biochar started to make a comeback as a binder in animal feed. For the first time in nearly a century, it is being fed to cattle as a regular supplement at approximately 1 percent by weight. While cattle and chicken farmers were the early adopters, the additive is now finding its way to sheep, goats, pigs, horses, rabbits, cats, dogs, and fish. Claudia Kammann, a soil ecologist at the Hochschule Geisenheim University in Germany, told us that everyone used to think the government would have to offer a program to get farmers to put biochar on their fields, but now a growing number of cattle farmers in Europe are adding it to winter feed and the cows are spreading it on the fields for free.

Weight gain for a dairy herd is only important for a heifers' first two years. After that, they try to maintain their girlish figures for the most part. So for dairy cows, and brood cows in the beef world, the efficient weight-gain angle probably doesn't matter. The livestock sector most interested in weight-gain efficiency is the part that finishes steers into burgers, fattens hogs, or raises meat poultry. Most mixed rations consumed by those animals contain binders such as bentonite clay or sodium bicarbonate (think baking soda) to deal with the pesticides and herbicides found on hay and other livestock feed, which can reduce weight-gain efficiency and cause health problems. A growing number of studies have found that feeding biochar to livestock not only increases weight gain but also improves health. The supplements also chelate antibiotic residues and target pathogens before they have a chance to build resistance.[1]

In 2013, a study conducted by the Johns Hopkins Center for a Livable Future recommended that all nontherapeutic use of antibiotics be first phased out and then banned.[2] The study criticized the US Food and Drug Administration's "voluntary plan" to phase out a mere two hundred antibiotics, out of the six hundred currently in use for animals, over the course of three years. Cows, pigs, poultry, and fish fed a steady diet of wide-ranging antibiotics excrete most of it in their manure and urine. When this manure gets used to fertilize crops, these antibiotics can reach wildlife and eventually humans, too. The risks that most concerned the Johns Hopkins researchers were the evolution of antibiotic-resistant bacteria and reduced antibiotic effectiveness due to excessive use. Some but not all antibiotics can be thermally destroyed via manure composting, but in colder climes this may not happen. And manure in pastures is not composted, allowing antibiotics to flow freely to the closest water body. While pyrolyzing manures eliminates antibiotics and most other toxins, keeping them out of livestock diets in the first place ought to be priority number one.

The reality is that the owners of large poultry ranches, hog farms, and feedlots are usually focused on producing the most meat at the least cost. Antibiotic feed additives stave off illnesses that crowded conditions and monotonous diets promote. This reality requires turning a blind eye not only to animal cruelty, but to the longer-term consequences of daily antibiotic intake such as contamination of air, water, and wildlife and antibiotic-resistant bacteria and viruses finding their way to clinics and hospitals around the world. The good news (well, sort of, because it still doesn't address whether animals should be treated this way) is that biochar provides most of the same benefits as prophylactically used antibiotics and several additional benefits that antibiotics cannot provide without all the associated drawbacks and dangers. Once excreted, biochar will also reduce odors from manures.

In Japan and Taiwan, feeding biochar to fish has been shown to be associated with weight gain and healthier fish. Biochar is coming into wider use for aquaculture and mariculture because it scavenges ammonia and the nitrates in fish tanks and enclosures. Whether fed to fish or broadcast to the enclosure, the carbon eventually sinks harmlessly to the bottom, possibly taking some nasty compounds with it. In a land-based operation, whether it goes to compost or gets dumped somewhere, it will have cascading benefits.

Animal Nutraceuticals

One of the most promising biochar markets is animal feed. Compared with the fertilizer market, the global animal feed market is enormous, and unlike the soil amendment market, is not seasonal: It weighs in at $325 billion and is expected to grow 4.6 percent per year through 2024. The animal feed–additives market is forecast to top $30 billion by 2025, just part of the larger $65 billion animal-health market that includes diagnostics, vaccines, and drugs.[3]

In one experiment, lactating goats received either bentonite clay or charcoal at 1 percent of a feed ration, slightly contaminated with aflatoxin, a sometimes lethal mold. After two weeks, toxic content in their milk had dropped 65.26 percent for the bentonite and 76 percent for the charcoal, without causing any change in the milk's nutrient composition.[4] A similar experiment with cows' milk showed reductions of 36 percent and 50 percent respectively. Mycotoxins most commonly reside on corn and cottonseed, but may also be found on soybean and distillers' grain. When consumed by livestock, some toxins can carry over into eggs, milk, and meat. High or long-term exposure can lead to health issues in humans. In the United States alone, the economic impact of aflatoxins on dairy is estimated to exceed $200 million per year. The potential for economic loss and liability for farmers is enormous.

Despite all these benefits, the use of plant-derived charcoal is not currently permitted as a feed additive in the United States for livestock expected to enter the human food chain; it's okay for cats but not cows. Researchers and biochar producers in Europe have been more fortunate. They've demonstrated multiple health and economic benefits to hogs, hens, and Holsteins and have won regulatory approval. Feeding charcoal to ailing animals is not a new idea. Even wild animals have been observed consuming it. Adding biochar to daily feed rations is relatively new, however.

To Big Pharma, biochar is a Big Threat. In 2015, American Addiction Centers warned:

> With those pockets comes a strong hand of political and legislative influence, to the tune of $2.9 billion between 1998 and 2014 on lobbying expenses, and $15 billion in campaign contributions between 2013 and 2014. For some people, this straddles the line between unethical and illegal. Even as the federal government issues massive fines to keep those companies in check, the industry itself is a significant contributor to the Food and Drug Administration's budget, leading to concerns of conflicts of interest and outright bribery.[5]

In 2014, *ProPublica* revealed the amount of money that US pharmaceutical companies have agreed to pay the US Department of Justice in fines and penalties for fraudulent market practices, such as promoting medications for uses that violate the federally approved standards. It exceeds $13 billion. Still, these companies recoup those losses in a matter of weeks. Even as they admit culpability and promise reform, none of that news makes it into television advertising, which accounts for 61.6 percent of companies' direct-to-consumer revenue. Marketing blitzes and millions of dollars thrown at prime-time commercial spots ensure they can keep therapeutic carbon off the shelves and out of the cattle shed if they so choose.

The global medication market was worth $1 trillion in 2014. Ten of the world's top pharmaceutical companies are projected to earn $20 billion from animal drug lines in 2018. They would most likely prefer not to have nutriceutical biochar (which can be locally produced) slicing into that pie.

Australian Doug Pow began feeding biochar to his pasture-raised herd five years ago to see if it might improve the quality of soil and hay from his pastures.[6] Biochar sweetened with molasses (⅓ kilogram or ¾ pound per day per animal) passed through the cows' rumens and landed atop the sun-baked soils of Pow's ranch. Dung beetles feasted on the carbon-rich excrement and helped convey it down through the soil profile as deep as 40 centimeters (16 inches). The combination of bovines and beetles gave Pow a no-cost delivery mechanism that has bettered the health of both cows and pasture. His pasture productivity went up 25 percent over three years. Everything improved across the board: carbon, oxygen, nitrogen, magnesium, calcium, potassium, silicon, sulfur. Many of his neighbors who gawked and jeered from the fence line are now following suit.

Stephen Joseph, a university professor who was dispatched to study Pow's farming practices for the government, calculated that Pow paid $0.06 per day for materials and $0.05 per day for labor to make his adjustments for the entire herd. He saved $1.10 per cow per day in hay costs and between $5,210 and $6,630 in fertilizer his fields no longer required. His soil carbon in his worst paddock went from 34 ppm to 86 ppm. With total expenses of $1,000, including labor, spread out over the course of three years, Pow produced an income of $49,500, compared with $12,250 per year under the previous system. And that was just in the first three years. The carbon keeps coming.

Pow is not only a cattle farmer; he also grows avocados. (He founded the avocado industry in Western Australia.) Avocados are a crop that needs irrigation and good soil. They have a low salt tolerance and a high requirement of oxygen and boron in the soil. Pow has eight hundred trees, the newer ones planted on tilled topsoil mixed with biochar.

"Avocado trees come from a volcanic andosol, extremely new soil derived from volcanic ash, different from any soil in the world," he said. "We are trying to chemically get the soil similar to that which they evolved in, and biochar assists that."

Recently Pow secured funding from the South West Catchments Council to measure how the avocado trees' growth was affected by biochar. To get precise measurements, he peeled back 1 foot (30 cm) of topsoil straight off the surface with a road grader, then came back and deep-ripped and rotary-hoed another foot. The removed topsoil was blended with 24 tons of biochar at different rates — 0, 5, 10, 20 percent — and graded back in rows. He planted his young trees on the soft soil and watered them equally.

"Everyone who comes to see this won't plant an avocado tree without biochar now — the difference is that convincing," he says. At 5 percent biochar the leaf area has doubled. The amount of chloride the trees take on from irrigation water is down 23 percent, and much less irrigation is needed. The new trees have better stem diameter and more flowers.

"We are trying to set up an even, natural system. They are huge trees; some will set 300–400 fruit a year when they are four years old. If every tree did that, that would be twenty-five tons per hectare and that's $150,000 a hectare. . . . Biochar is not expensive and it's not like a fertilizer that is gone each year, it's in there for thousands of years."[7]

How Are Your Cows Sleeping?

Apart from ranchers, we suspect few people stay up at night pondering how their cows are sleeping. In the unlikely event that you are one of those people, allow us to shed a little light on appropriate bedding. The main criteria for selecting stuff to feather the bovine nest combine health-related concerns — it needs to be comfy, dry, absorptive, and should inhibit bacterial growth — and economic soundness — it should be cheap and plentiful, and the less labor is required to get it in and out, the better. Common inorganic bedding products include sand, which is becoming more expensive and can be a nuisance to handle; limestone screenings; and gypsum, which comes from recycled drywall and can be hazardous. Porous carbon would at first glance seem a better choice for managing pathogens than sand, gravel, sawdust, straw, shavings, or dried manure.

While mucking stalls could never be classified as glamorous, it never seemed hazardous until alerts were issued by research laboratories at Cornell[8] and Penn State.[9] Some dairies use a gypsum bedding additive to reduce odors, absorb ammonia, and suppress pathogens. Now we discover the same product caused mastitis and other nasty ailments in the cows, and lethal effects to farmworkers and their families. The researchers described the deaths of farm infants and toddlers exposed to gypsum fumes, teenagers and young adults found dead in enclosed spaces, and fatalities of 22 percent of would-be rescuers who attempted to enter the spaces and resuscitate the victims.

Although biochar may be restricted as feed to livestock in some regions, nowhere is it prohibited as bedding material in barns and sheds. The benefits are compelling: reduced odors, nitrogen retention, and increased time between bedding material changes. And when the litter gets composted and becomes fertilizer, it does the whole climate change thing. The cost of most livestock bedding materials is relatively low in comparison with the current price of biochar, but pricing for both can fluctuate widely, depending on the season, proximity to producers, alternative uses of source material, and other variables. If you make biochar yourself, it can cost anywhere from nothing to a few hundred dollars per ton, depending on what you have to work with.

The inoculated deep litter system (IDLS) is a great example of using biochar in bedding. This system has been in use South Korea, where it was developed, and has been used at more than fifty piggeries in the United States since 2009, as well as in countries such as Vietnam, Ireland, and the Philippines. Building upon a 6-inch (15 cm) layer of biochar, IDLS uses logs, green waste, and sawdust inoculated with indigenous microorganisms. It never needs to be cleaned out; saves water and labor; doesn't generate additional waste; requires less land for manure management; reduces odors, flies, and rodents; reduces the need for vaccines or antibiotics; and puts less stress on animals.[10] Deep litter systems have moved from pigs to poultry and will likely find their way to other livestock in the not-distant future.

Leafy Greenbacks

The world population is currently growing at a rate of 1.14 percent per year, or about an additional 220,000 persons per day. And yet despite this growth, in a little more than a half century, the global per capita supply of protein has increased by about one-third, from 61 grams per day in 1961 to 81 grams per day (~2 to 3 ounces) in 2014.

There is, however, a marked difference in protein supplies across the world. This difference between regions is more stark when you measure by protein instead of by calories. The per capita supply of protein in North America was 60 percent higher than in Africa in 2014. In terms of calories, the difference was 40 percent.

It's not just the quantity of protein that has changed but also the composition of the sources. The world gets much more from meat, dairy, and seafood today than it did fifty years ago. As incomes increase, people tend to eat more animal-based protein than plant-based protein. Typically, animal-based proteins are defined as "complete proteins," meaning they contain all the amino acid building blocks we need for proper nutrition. Proteins found in crops — with some exceptions such as pulses, some nuts, and seeds — are often of poorer quality and lacking in some essential amino acids. The increase in animal protein in diets also leads to a larger, indirect, increase in demand for feed because of the inefficiency of converting from plant to animal protein. From 1 acre of good land you can expect 2,500 pounds (1,136 kg) of soybeans, 11,500 pounds (5,250 kg) of maize,[11] or, alternatively, 1.6 grass-fed steers that would gain 525 pounds (239 kg) from spring to fall.[12] For this reason, most cattle are raised on grain — it simply takes less land. The same reasoning is why cereal grains, not meat, built so many of the monumental civilizations of antiquity — Sumer, Egypt, China, Greece, Rome, the Aztecs, and the Inca.

In 2030, cereals and oilseeds will likely still be the major suppliers of our protein, as they are today. If animal protein were to supply two-thirds of what we demand by then, 696 million tons of protein from cereal and oilseed crops will be produced for animals. Shifting the human diet to more plant and less animal protein is an optimistic but improbable scenario, but even if that happened, more feed proteins will be demanded than can be produced on the land available. Population is finally, somewhat belatedly, arriving at Malthus's predictions.

These limits would be greatly aggravated if lands producing cereal and oilseed crops were diverted to produce biomass energy crops. We have had a taste of that with corn used for ethanol. In the United States alone 90 million acres (364,217 km^2) of arable land are devoted to growing corn; a third or more is used for ethanol production, and another third is for animal consumption.

Consider, too, that by 2030, the chemical industry is expected to replace fossil-based feedstocks in chemical production by up to 20 percent. Some

40 million–65 million tons of plant proteins will be required by that indus-try for chemical feedstocks.

Protein Presses

Jonathan Male, who heads up a joint task force for the US Department of Energy's Bioenergy Technologies Office (BETO), in partnership with Oak Ridge National Laboratory and other federal agencies, reported in 2016 that no single crop, looked at in isolation, can make the tricky business-and-environment case for bioenergy on its own, but put them together and you have a food-and-energy revolution. The Oak Ridge National Labora-tory pegged sustainable production potential at 1 billion dry tons of biomass resources — from agricultural and food waste, forestry, and algal biomass — annually by 2040, in the United States alone.[13]

That seems nearly impossible if, like most environmentalists, we con-tinue to look at food and fuel as competitors for the same land. However, we should ask whether that is a misleading dichotomy.

In 1773 the French scientist Hilaire Marin Rouelle mashed several species of leaves in a beaker, then strained the juice and applied heat. A green curd formed and floated to the top as the leaf juice approached boil-ing. Rouelle reduced the residue and analyzed its contents, then published a paper on the subject. The paper was of interest only to a small number of academic researchers.

During World War II, when German U-boats took control of the North Atlantic shipping lanes and began torpedoing supply ships heading to England, Rouelle's work suddenly gained new meaning. With rationing stretched to its limits, the British government began a crash program to study alternative foods. Food scientists began extracting protein from the green leaves of alfalfa, barley, wheat, mustard, algae, and other plants. Similar work was carried out in Switzerland by the French cooperative France Lucerne. After the war, retired executives from France Lucerne established L'Association pour la Pro-motion des Extraits Foliaires en Nutrition — that remains active today — to apply leaf concentrate to combat malnutrition in Africa, Latin America, and elsewhere, including nutritional supplements for people with AIDS.

These programs showed that green leaves not typically consumed by humans can be potent sources of food when refined for their high-quality amino acids, vitamins, and minerals. One hundred grams (3.5 ounces) of dried alfalfa leaf concentrate has about twice the protein of beefsteak, dried whole milk, or scrambled eggs. It also has twenty to fifty times the iron, up to one hundred times the calcium, and ten to twelve times the vitamin A. Leaves from the African moringa tree are even more nutrient-dense. In fact, a great many leaf concentrates are richer in these essential nutrients than many of the more commonly available foods.

Albert first met David Kennedy in the 1970s, when he showed up at The Farm to eagerly tell its residents about alfalfa juice. Albert tasted some, politely, and then quietly slipped outside to expectorate. It was bitter and unpleasant, and the memory lingered. Kennedy was a member of the Flat Rock Community near Murfreesboro and active in the Cumberland-Green Bioregional Council. At each of the quarterly gatherings, Albert recalls, "He would bring forth his latest discoveries and show us all how to take edible leaves, juice them, and then curdle the juice with vinegar to create a green tofu paste he called 'leaf protein concentrate.'"

Kennedy later wrote in his book, *Leaf for Life*: "In addition to the high levels of nutrients, the nutrient bioavailability is very good compared to other plant-based foods. This is because the fiber from the cell walls has been separated and removed during the process. This combination of high nutrient levels and good bioavailability makes leaf concentrate quite effective in combating malnutrition. Not only are the fiber and water separated during fractionation, the process also removes hydrocyanic acid, free oxalic acid, and nitrates that limit the usefulness of many leaf crops. Because the juice is heated to the boiling point, *E. coli* and most other pathogens are killed."

In 2015, an Indonesian PhD student, Yessie Widya Sari, looked at some of the most readily available food-processing wastes.[14] She extracted protein from rapeseed meal, soybean hull, soybean meal, whole soybean mill run, sunflower meal, palm kernel meal, microalgae and microalgae meal (*Chlorella* sp.), wheat middling, wheat gluten, barley grain, barley mill run, barley rootlets, malt by-products, ryegrass, and sugar beet pulp. Using a relatively simple process of a warm alkali bath for twenty-four hours followed by centrifuge and cooling, she extracted and analyzed a protein residue. As might be expected, different biomass types had a variety of protein concentrations, ranging from 8.6 percent for ryegrass to 77.5 percent for wheat gluten. The important point is the products

Sari tested are nearly always regarded today as industrial wastes, not potential foods.

Whether it is economical to extract protein from leaves, meals, and grasses is another story, but these organic materials must be considered within the carbon cascade framework. Sari learned that palm kernel meal protein requires €3,069 (~$3,700) per ton to process, which disfavors palm kernel meal as a protein resource. However, combining protein and energy production (pyrolyzing the nonprotein residue) returns revenue of €2,728 (~$3,300) per ton. Using the residues as cellulosic fiber for feed is even better for palm kernel meal, giving additional revenue of €3,131 (~$3,800) per ton, already enough to cover processing cost, before any revenue is made from the protein product. Rapeseed also has its highest revenue when protein and feed production are combined. However, wheat middling can be better used for protein and energy. Each form of biomass must be looked at for its optimized value, but generally speaking those highest in lignin are best for fuel and those highest in cellulose are best for animal feeds. All are suitable as nutritional supplements for humans if properly processed.

At the University of Hawai'i at Mānoa, bana grass (*Pennisetum purpureum*) was hand-harvested at four months, shredded, and fed into a screw-press. The extracted juice was used for cultivating an edible fungus, *Rhizopus oligosporus* (traditionally used in making tempeh), for aquaculture feed supplement. The bana grass juice increased the fungal growth rate by 16 percent.[15]

A special case for costing must be made for microalgae and their meals because of greater production cost, higher processing cost, and low protein extraction yield, with less than 10 percent protein extracted at room temperature. And yet, algae offers unique products that could more than pay these added costs, most notably biodegradable plastics and biocrude petroleum.

CHAPTER 10

Manure Happens

Not many wild animals foul their own nests, but we humans have managed to do it on a grand scale. Some places and people do almost nothing to regulate or contain their excrement and give little thought to its future impact. The result, predictably, is damage to health, not merely to humans, but to terrestrial and aquatic ecosystems and all our relations.

The very idea that excreta are waste with no useful purpose is a misconception. Chinese farmers have been composting human and animal excreta for a few thousand years, and Japan introduced the practice for agriculture in the twelfth century. Western fecal-phobia is the starting point for conventional approaches to sanitation, particularly flush-and-discharge. The energy required and the pollution problems that result from this choice are enormous. In nature there is no waste. All living things and their excretions are food for something else.

Pyrolysis and humanure can be a winning pair in composting toilets, wastewater treatment plants, and everything in between. Reclaiming water after use is a design consideration of growing importance. Waterless sanitation systems are first responders to our need to conserve potable water. Hastily constructed dry toilets, however, may not be equipped to deal with the complexities of pathogen destruction and excreted genetically modified organisms, glyphosate, and pharmaceuticals. Typically in such systems, different types of carbon, such as sawdust, paper, wood chips, and leaves, are tossed in to help reduce odors and facilitate decomposition. While these may speed decomposition, they do not remove all the hazards. Adding biochar captures nutrients for later use, arrests odors, and further speeds the decomposition process. It can also sequester and decay some hazardous elements, but not all.

A number of environmental conditions will speed up or slow down the time it takes a pathogen to die. The primary die-off accelerants are temperature, moisture, nutrients, other organisms, sunlight, and acidity (pH). Each of these conditions can vary naturally between, for example, wet and dry seasons, or artificially, by addition of lime, for instance. The greater the numbers of disease organisms present, the longer it takes for all the organisms to die.

Biochar, by creating conditions most favorable to aerobic bacterial health, can speed pathogen destruction. As we've said earlier, a temperature above 60°C (140°F) will result in near-instant kill for most pathogens excreted in feces. Temperatures in the range of 50–60°C (122–140°F) result in no growth for bacteria and death within typically thirty minutes. Biochar can also help by regulating moisture in the compost. Increasing temperature *and* decreasing moisture will produce a faster die-off than the manipulation of only one factor.

Of the methods of pathogen destruction, high-temperature composting is best but not perfect — parts of the compost heap may not reach the required temperature, so some pathogens may survive. Bacteria, viruses, and protozoa usually take several months to die completely. Eggs of the species *Ascaris lumbricoides*, the common roundworm (which infects an estimated 20 percent of the world's population), can survive for years.

Pathogen death rates in soil are increased by dryness and sunlight — in sandy, sunny soils, within two weeks. In Guatemala, where *Ascaris* infestation is present in up to 50 percent of the population, thousands of eggs per gram have been found in dry compost toilets. Storage and dehydration in the toilet's processing vault followed by open-air dehydration reduced the egg numbers to zero after six months of sun drying. Biochar can change soil properties to be more sandlike, so adding it to clay soils will hasten *Ascaris* die-off.

Waste stabilization ponds — natural or constructed wetlands with both aerobic reed beds and anaerobic settling lagoons — are faster and more effective in destroying protozoa and helminths (parasitic worms), including *Ascaris* eggs, but bacteria and viruses may still be present in the water that leaves the ponds. Biochar filtration at the end of a wetland treatment process, followed by soil amendment with "charged" char, is a more complete solution and a compelling carbon cascade.

Many households rely upon septic tanks and leach fields, which can benefit from including biochar in construction and operation. In the grease trap leading to the settling tank and in the tank, biochar's favorable habitat for microbes accelerates the liquefaction of solids. Beyond the tank lies the leach field. A three-bedroom home on a septic system requires roughly a 450-foot linear run of perforated pipe in gravel for sandy soils and possibly double that for clay soils. On average, the cost of creating or repairing gravel leach beds ranges from $5 to $15 per linear foot, or between $2,250 and $6,750 for a 450-foot distance. Displacing gravel with biochar could help reduce both size and cost of installing a field. Biochar is also superior to gravel in the biological activity it encourages and in odor removal.

If a homeowner chose to add a 1-inch layer of biochar, for example, she would need roughly 2 cubic yards for a 450-foot area, which would cost between $20 and $60. Or, if she chose to replace the soil almost completely with 18 inches of biochar rather than gravel, she would need 36 cubic yards, costing between $390 to $1,100 at present prices.[1] She could reduce the cost by making char onsite using some of the biomass uprooted to make way for the leach field. Depending on the amount used, a homeowner could sequester anywhere from 2 tons to 4 tons. In areas where homes do not have much land or topsoil is sitting atop rock, biochar-based leach fields may be the only option.

Greenport, a Swiss company that produces and rents mobile dry toilets, is taking a slightly different tack. When rental units are serviced, the human waste is trucked to a carbonizer. Toxins like human pathogens, metals, and pharmaceuticals are removed, while carbon and some beneficial nutrients are preserved. The biochar is sent to a local zoo for landscaping. Carbonizing sewage waste not only offers an elegant solution with many possible new products, but decarbonizes an industry that now contributes nearly 3 percent of global greenhouse pollution.

Adventures in Sludge

Because many animal farmers operate as concentrated animal feeding operations (CAFOs), managing vast amounts of manure has become big business. The most common manure management processes include anaerobic digestion, slurries or lagoons, composting, and more recently, thermochemical conversion. Biochar can play a mitigating role in each of these different systems.

How Far Is Away?

New York City is the beating heart of global finance, a cultural behemoth, and home to more than 8.5 million people who create an enormous amount of poo. Some of this expelled waste has been causing a major stink 900 miles (1,448 km) away, in Alabama. Residents in and near Birmingham have been in uproar over sewage that is transported by train and truck from New York and New Jersey to be dumped in poorer rural areas. The treated sewage — euphemistically referred to in the industry as "biosolids" — has plagued residents with a terrible stench, flies, and concerns that spilled sludge has leaked into waterways. According to a 2018 article in the *Guardian*:[2]

"On a hot day, the odor and flies are horrific," said Charles Nix, mayor of West Jefferson, a town near the landfill that accepts the waste. "It's better in winter time but if the wind blows in the wrong direction you get the smell. It's like dead, rotting animals.

"If you get close to the trucks the liquid would blow off on to your windshield and fill your car with a stink. It spilled out onto the road. Some people were saying they just wanted to move away, they were so miserable."

Within the past few months members of the NYC Office of Environmental Protection responsible for managing biosolids have been inquiring about carbonization. There is hope that at least one never-ending pile of poo can be converted from stink to sink.

Anaerobic digestion enlists the help of microbes to break down different organic materials into three by-products — biogas, digestate, and effluent. Biogas in the form of methane can be converted into various types of renewable energy or converted into transportation fuels and can be critical to making anaerobic digestion systems financially viable. Manures on their own do not produce sufficient methane to justify the high cost of digesters, so farmers often accept other organic materials for a tipping fee to increase energetic value and methane output.

Achieving consistent and high-level methane is time consuming and often beyond the typical skill set and interest of farmers. This has inhibited the adoption of methane digesters in many places, although in land-constrained areas or areas with significant regulations on organics disposal, such as Europe, the practice is far more common.

Argonne National Laboratory discovered that adding small amounts of biochar to anaerobic digesters can boost both the quantity and the quality of methane. This process led to municipalities being able to reduce contaminants from sewage sludge, and that's led to pipeline-quality methane for power and transportation fuel. Likewise for small farmers, adding small amounts of biochar to manure digesters will improve both yield and quality of the biogas. It also reduces the lag time for methane production, cuts odors such as ammonia or the rotten-egg smell of hydrogen sulfide, and inhibits corrosion of equipment.[3] All of these benefits also make these systems more economically viable.

Farmers are still left, however, with a pile of organic materials once the energy value is depleted; 90–95 percent is left behind as both a fibrous solid called digestate and an effluent. While there are some markets for the digestate, they're often limited to local farmers that will dry it for bedding or apply it to fields. Carbonizing it can reduce volume, retain some nutrients, and produce biochar. The heat generated during carbonization can also be used to dry the digestate.

Like hog farms and poultry barns, large-scale aquaculture tanks or lagoons also produce effluent rich in nitrogen, phosphorus, and suspended solids. Filtering fish wastewater with biochar — or better still, feeding the biochar to the fish and letting them inoculate the water — can convert the biochar into first a water filter and then a slow-release fertilizer. Instead of incurring costs for effluent management, fish farmers can monetize their excess nutrients to produce slow-release fertilizers.

Slurry

Imagine a swimming pool filled with fecal matter sloshing around in urine with a cover on top. That, in a nutshell, is a manure slurry, lagoon, or pit. Livestock producers use these holding areas to store manures, used bedding materials, and stall-cleaning rinse waters until they can be composted or applied to fields; in colder climates, this can mean six months or more. The slurry is pumped out of the pits and into the fields to prevent it from moving off the fields into neighboring water bodies and to reduce odors. If slurries overflow after heavy rains, they can contaminate soil and groundwater with pathogens (*E. coli*, *Listeria*, *Giardia*), antibiotics, heavy metals, and hormones. That rotten-egg smell from hydrogen sulfide wafts far and wide. So, too, does ammonia, causing harm to all manner of flora and fauna. Oh, and by the way, slurries also exhale prodigious amounts of carbon dioxide and methane.

Using biochar as a floating cover on lagoons has been shown both to reduce ammonia and to adsorb certain nutrients, including nitrogen, calcium, magnesium, sodium, iron, aluminum, silicon, and in some cases phosphorus. It was also found to reduce odors.[4] Chinese researchers have shown that woody biochar with high surface area and small particle size (0.25 mm or 0.02 inch) can adsorb high amounts of ammonium in pig slurries (73 percent) and in digested pig manure (60 percent).[5] Other researchers found that applying woody biochar atop the slurry can reduce ammonia related to dairy manure slurry by 96 percent.[6]

With housing developments encroaching on traditional farming areas, odor control has become a huge topic. This aspect alone might spur livestock farmers to investigate biochar as an option to avoid litigation from their new neighbors.

Composting

Composting manure on livestock operations is a common practice and generates valuable organic fertilizers. However, it is also responsible for between 30 percent and 50 percent of the annual global production of agricultural N_2O emissions, and a good portion of methane as well. Not only are those two greenhouse gases emitted during composting, but the compost continues to give off CO_2 and methane after being land-applied, and the carbon it adds to the soil stimulates the soil biota to make more N_2O. Blending biochar with manure compost can lower these emissions, reduce odors, shorten composting time, increase nitrogen retention, stabilize acidity, and burn off more harmful pathogens.

Heavy metals are now pervasive in soils, often the result of overzealous application of synthetic fertilizers, fungicides, or mineral feed additives.

The metals are taken up by plants, ingested by animals (and humans), and then excreted. Many of these, like copper and zinc, are not eliminated through composting but instead get concentrated. When farmers spread composted manure, they can make the heavy metal problem worse. Biochar added to a well-rounded fungal compost makes sense. It can immobilize heavy metals, making them unavailable to plants.[7]

Thermochemical Conversion

Where there is insufficient land on which to spread the digestate, slurry, or manure, farmers and herdsmen may need to explore other manure management options. Thermochemically converting manure through either slow pyrolysis or gasification is one alternative. Manure volume can be reduced significantly, from 42 to 63 percent, depending on production temperatures. While much of the nitrogen may be volatilized (this is also dependent on temperature), many other compounds can be preserved as plant-available nutrients. Carbonizing high-moisture-content biomass can be challenging, but new technologies are available that utilize the heat generated during pyrolysis to dewater manure or digestate.

Current systems to carbonize manures can cost in excess of $1 million, making them viable only for larger operations. With some areas starting to regulate methane emissions, many more feedstocks are going to get pyrolyzed. Companies like Jiangsu Benenv (for "benefit the environment") of Yixing, China, have foreseen this opportunity and gone into production of low-cost, small-volume, high-efficiency, mobile, modular, quality sludge carbonizers, screw presses (useful for extracting leaf protein prior to carbonization), and power co-generators. Where others see only a receding tide, they look to the horizon and see an incoming tsunami of carbon opportunities.

Carbonizing Sludge

Arriving from rivers and coastal discharges into the world's oceans, sludge would make a wonderful fertilizer for seaweed if it weren't contaminated with plastics, solvents, pesticides, beauty care products, cleaning products, and everything else we flush away. What goes down the drains in hospitals can be laden with pharmaceuticals, including birth control hormone regulators, biohazards, and radionuclides. Some pollutants can be captured by settling lagoons, filtration, ultraviolet light, or other stabilization processes, but many survive even those systems.

Although some countries have lax regulations when it comes to land application of biosolids, standards in Europe, North America, and elsewhere have strict heavy metal thresholds and nutrient loading rates. Even there, certain materials found in biosolids may be unregulated, and in many cases untracked, including dioxins, pesticides, microbeads, microfibers, and more. Many of these contaminants persist throughout a modern wastewater treatment cycle and sludge stabilization process. Once applied to soils, they can find their way into our food and water.

Triclosan (TCS) and triclocarban (TCC) are antimicrobials commonly found in toothpaste, soap, shampoo, mouthwash, and pesticides. Nonylphenol (NP) is a chemical compound frequently found in laundry and liquid detergents and is an endocrine disruptor. TCS, TCC, and NP are not currently regulated and so they are commonly found in sewage sludge. Pyrolysis can eliminate these if the correct temperatures and residence times are used.[8] Another persistent micropollutant is estrogen from medicines such as birth control and hormone treatments, as well as from livestock waste. Estrogen from anaerobically digested biosolids has been linked to intersexing of aquatic animals. Pyrolysis at temperatures above 400°C (752°F) reduces estrogen by more than 95 percent.[9]

Landfilling biosolids is costly and comes with a heavy carbon footprint from transportation and released methane. One small treatment plant in upstate New York spends $300,000 to landfill 4,500 tons of biosolids every year. Carbonization would reduce the volume of that waste to less than half. Even if no markets could be found for the biochar, the cost savings would be $150,000 annually. The high moisture content and odor create challenges for landfills. They may also leach out pathogens, chemicals, and heavy metals whenever the water table rises.

In Europe, roughly 21 percent of treatment plants dispose of their sludge via incineration, which can reduce its volume by 70 to 95 percent.[10] Sludge ash is sometimes used as a filler for cement, in road construction, or as daily landfill cover. Although energy is required during the start-up phase of incineration, the heat produced becomes largely self-sustaining for most of the process. (The city of Nashville, Tennessee, found it could heat buildings in the downtown area with steam produced by its municipal incinerator until dangerous levels of dioxin, a toxic by-product of burnt plastic, started appearing in residential neighborhoods and schoolyards.) That said, although sludge incineration has a lower carbon footprint than landfilling, valuable nutrients can be lost. Incinerators also consume significant amounts of water — 758 gallons (2,869 liters) per ton of processed weight (2.73 liters per kg).[11]

By-products of thermal conversion of sewage sludge include synthetic gas (syngas), bio-oil, and pyrolysate. Carbonized biosolids have relatively low levels of carbon, normally 10–50 percent depending on production temperatures, as compared with woody feedstocks or crop residues. As such they may or may not be considered biochar depending on the classification standard. To be considered biochar under the European Biochar Certificate a minimum of 50 percent carbon is required. The International Biochar Initiative (IBI) Biochar Standards classifies 10–30 percent carbon as a class 3 biochar.

Phosphorus in biosolids char is a plant-available variety that works as a slow-release fertilizer. While a small portion of nitrogen is retained, it is insoluble in the short term and not available for plants.[12] Without treatment, soil acidification from fertilizer and acid rain can have a major impact on agricultural productivity and can extend into subsoil layers, creating poor conditions for nutrient transfer and plant root development. Carbonization with different production parameters, feedstock blends, and postprocessing technologies can create biochars with a range of acidity from a low pH near 5 to a high (very alkaline) of more than 10. Carbonized sludge can be applied as a liming agent to deacidify damaged soils. For these reasons it should not be surprising to find a 50:50 mix of pyrolyzed biosolids and wood chips is a class 2 (mid-range) fertilizer due to phosphorus and magnesium content; a class 4 (the highest class) liming agent; and a class 1 (the lowest class) carbon source using the IBI biochar classification system.[13]

In controlled tests conducted in Australia, fertilizer-enhanced sludge char showed an increase of 167 percent in cherry tomato production compared with a nonfertilized control, 27 percent over a fertilized but unpyrolyzed control, and 60 percent over unfertilized pyrolysates. Assays for this particular product showed the bioavailability of heavy metals to be well below permitted maximums except for cadmium, which was close to the allowable limit for Australia.[14]

Biomass Controls, a production technology company based in Connecticut, sells an off-the-shelf, community-scale (ten-thousand-person) pyrolysis plant that processes human sewage or other high-moisture-content feedstocks. In collaboration with the Bill & Melinda Gates Foundation's Reinvent the Toilet Challenge, Biomass Controls has been testing units in parts of India where human waste previously went untreated. They've shipped units to remote Alaskan islands and are looking at servicing displaced refugees and other marginalized communities. Super Stone Clean of Japan, Jiangsu Benenv of China, and Pyreg of Germany make carbonizers especially designed for sewage sludge. These range in size from Benenv's factory-floor-sized Super Rotary Dryer and Carbonizer down to Super Stone Clean's series Z devices that travel on a small truck.

While metals are largely sequestered in biochar's pore structure, regulations control the level of metals in any soil amendment, especially if it is used for growing food. This can significantly limit where sludge chars can be applied. Interestingly, some golf course managers prefer the heavier weight of sludge chars as compared with other lighter chars because they are easier to apply to the greens. They have found biochar reduces the need for irrigation—a substantial expense for golf courses in dry climates. Biosolids have also been used for reclamation projects because they speed revegetation and mitigate erosion and leaching. Fly ash and lime are often added to improve pH, but recent studies have shown that this use of biosolids can degrade water quality and increase leachability of heavy metals. Sludge chars and limed sludge chars effectively reduce acid and basic dyes and phenols, and can be safely used for filtration and remediation in many settings. Stunningly, it has been found that gold and silver can also be inexpensively removed from biosolids by carbonization and a second-stage acid treatment of the biochar.

Combining sludge with concrete or sintering with clay is a viable method of further stabilization of heavy metals. Japan has created a growing market for dewatered sludge as an energy source and additive for the production of Portland cement. Researchers in Brazil have found that small amounts of sewage sludge could effectively be incorporated into roof tiles. Taiwanese researchers baked sewage sludge at 250°C (480°F) and found that using up to 10 percent torrified sludge could still produce

Table 10.1. Carbon Math: Sludge

World wastewater production	108 mi³ [450 km³]
Weight of sludge [720.8 kg/m³]	324 Gt/yr
Dewatered sludge at 3% solids	9.7 Gt/yr
Convert 20% to class 3 biochar	1.9 Gt/yr
Weight after carbonization [25%]	486 Mt/yr
With 30% C content	146 MtC/yr
CO_2-e drawdown potential	(535 $MtCO_2$-e/yr)
CO_2-e avoided from transportation, landfilling, or land application	unknown

Note: Potentials for atmospheric removal are in parentheses. Mt = megatons. The degree of substitution is based upon the physical properties of the substance being displaced and the authors' estimate of market penetration potential.

good-quality bricks. Charring sludge not only reduces the volume of an expensive, ubiquitous problem but can also provide construction materials with improved qualities of fracture resistance, humidity control, and insulation. A typical cascade might be pyrolysis for energy, emissions, and volume reduction; pyrolysates used as filtration media at the facility; and finally, long-term encapsulation in building materials.

Global wastewater production is 450 km³ (108 miles³) per year.[15] At 720.8 kg per cubic meter (45 pounds per cubic foot), that sludge weighs 324 gigatons. Dewatered to 3 percent biosolids and carbonized with a yield of 25 percent biochar by weight, it would more than supply the demand for the kind of charoset, biochar-cements, and asphalts we described earlier. If 20 percent of the world's wastewater — much of it too contaminated with heavy metals, pharmaceuticals, fertility hormones, and other pollutants to be suitable for soil — were carbonized to class 3 biochar with 30 percent recalcitrant carbon content (146 Mt), 535 million tons per year of CO_2-e could be withdrawn from the carbon cycle and entombed in buildings, roads, airports, and dams.

Carbonizing this eminently renewable supply of underutilized organic matter could reduce many, if not most, of the negative environmental impacts associated with biosolids: odors, toxic leaching, greenhouse emissions, and more. As the cost of organics disposal rises, due to growing population and shrinking land availability, and the cost of carbonization comes down thanks to cascade synergies, the curves will cross and carbonizing biosolids will suddenly be both the most affordable choice and also the best for the planet.

Biogas Cowboys

In 2011, while finishing his master of architecture degree at the University of Tasmania — his honors thesis coined the word *permatecture* to describe integration of permaculture, architecture, and urban problems such as absorbing refugees and the homeless — Stuart Muir Wilson began thinking about methane. He eventually envisioned a way the forty-five hundred dairy farms in his state of Victoria could boost their bottom lines by converting to organic production and reusing effluent.

Biogas digesters are not new. In 1859, a leper colony in Bombay built the first anaerobic digester, and by the late 1870s England was powering gaslights from their sewers. The Chinese and Indian governments have distributed small biogas generators for household cooking and lighting for the past half century. Other projects receive financial support through the United Nations Clean Development Mechanism and the Global Environment Facility.

One problem with many methane digesters, especially small, on-farm units, is that they leak. We asked Wilson about this. We said we could agree that biogas is preferable to fossil energy because the energy is derived from recently grown plant material first digested through the stomachs of humans or livestock. That takes the carbon back from the atmosphere. But if you're manufacturing methane, and some significant portion goes to the atmosphere, that may not be much of an improvement. Moreover, if you are burning the methane for fuel, as most post-digester systems do, you're introducing another problem.

During the reaction to produce energy, one molecule of methane combines with two oxygen molecules to form one CO_2 molecule and two water molecules, which are usually given off as steam or water vapor. This process is taking one greenhouse gas (methane) and producing two greenhouse gases (CO_2 and water vapor). This conversion happens daily in millions of home kitchens in China and India and yet is still portrayed as climate-friendly.

Burning plants in the field is bad, but it only produces one greenhouse gas molecule per atom of carbon.

"Biogas digesters should be banned," we said. Wilson laughed and answered with one word: "Algae."

Wilson readily agreed that the way biogas is produced in many places, particularly by all the small producers in India and China, is bad for the atmosphere. But he had a carbon cascade solution. He grows algae in a closed container and draws in bubbles of methane gas from his biodigester, which he stimulates to greater microbial fermentation using biochar. The algae absorbs the carbon and yields hydrogen for fuel. The algae can then be harvested for food or pyrolyzed for more energy and to produce biochar. No greenhouse gases escape. He has branded his creation Bullock-Proof Energy Systems.

The discussion about methane is important because methane is a nasty piece of work as carbon molecules go. It could lead to a tipping point that would push Earth out of the Goldilocks zone (not too hot, not too cold) and closer to the blast-furnace climate of Venus. Enormous amounts of methane have been trapped on the frozen ocean floor and in permafrost by millions of years of plant material decomposing with no oxygen. Global warming has begun to thaw and release those stores. The release of just 10 percent would trigger an impact equivalent to ten times the current, and already catastrophic, greenhouse effects of carbon dioxide.[16] Another potential source of atmospheric methane may be the so-called fugitive emissions from exploration, production, and pipeline or rail car distribution of unconventional fuels such as shale gas and fracked natural gas.

The methane debate has grown heated in recent years in part because of blame heaped upon the livestock industry for methane emissions from both enteric sources (e.g., cow burps) and manure management. If Wilson's biogas-to-algae system could capture a significant proportion of that methane and convert it to clean energy, the soil-building benefits of raising cattle and poultry would transform animal husbandry from a carbon source to a carbon sink. Unfortunately, stand-alone biogas digesters may never be cost-competitive with renewables such as wind and solar PV for power generation. That cow has already left the barn. What they might produce competitively is something wind and solar can't: liquid fuels.

Yet it is not an either/or choice — electricity from a pyrolysis kiln or biogas from an anaerobic digester. They complement each other. Gas from the digester can fire up the kiln and biochar produced from the kiln can improve the quality and quantity of methane being generated by anaerobic digestion. Wilson's design adds even more cascades. The biodigester, optimized by biochar, stimulates the growth of algae. The algae, doubling in mass almost daily, can be harvested continuously, and its biomass can be converted to protein supplements, probiotic nutraceuticals, cosmeceuticals, and other products before sending fibrous residues to be dried and pyrolyzed for distillates, energy, and biochar.

Microalgae have much faster growth rates than terrestrial crops, because the cells grow in aqueous suspension where they have more efficient access to water, carbon dioxide, and dissolved nutrients. The per-unit-area yield of oil from algae is estimated to be between 2,000 and 20,000 gallons per acre, per year (4.6–18.4 l/m-yr). This is three to thirty times greater than the next best crop, Chinese tallow (699 gallons/2,646 liters). Algae can produce up to 60 percent of their biomass in the form of oil. Refined algal oil is already being used as a substitute for jet fuel in experiments by commercial airlines.

The farmer benefits from environmentally responsible waste removal; power; distillates; a range of valuable foods, fibers, and feeds; and biochar. The climate benefits from reduction of methane and storage of carbon in its recalcitrant form. Early trials of Wilson's prototype system suggest a return on investment of around 50 percent, with the capitalization costs recovered in two years.

CHAPTER 11

Blue-Green Revolution

In 1658, a Japanese officer arrived at a little inn. The innkeeper, Minoya Tarozaemon, ceremoniously received him and offered a traditional jelly dish as dinner, which was prepared by boiling seaweed (*Gelidium* sp.). After dinner, the surplus jelly was thrown outdoors, froze during the night, thawed the next day, and dried in the sun. Tarozaemon found this soft substance and boiled it in water. When it set, he discovered a whiter jelly than the original one.

In this way, according to Japanese lore, was born agar, or *kanten*, meaning "cold weather." In China it is *dongfen* or "frozen powder." The Malayan term is *agar-agar*, which referred to jellies of the seaweeds in the East Indies. Europeans who were living there learned to use this product for making fruit jellies, and took it to Europe. There scientists discovered that it made a useful culture medium. In 1882, it was used to identify tuberculosis. It has been in biology research laboratories ever since.

Seaweed is a cascade opportunity waiting to be seized — from red forests of blue carbon producing food, fiber, bio-oils, fuel, and flavorful gourmet "sea"soning to carbonization of the remains as sea-char, concrete for seawalls, and green charcoal. Seaweed is already big business for much of the world, but the seaweed industry seems as yet oblivious to the benefits carbon is waiting to bestow when they are ready to cascade. As good as biochar is at scavenging nitrates and preventing them from reaching the sea, nothing is as good at capturing them as photosynthesis in the form of sea flora and coastal wetlands, once the nutrients get there.

A textbook example can be seen in the 17,000-square-mile (44,030 km^2) Minnesota River basin. Surveying that land area from satellites in 2018, scientists determined wetlands are five times more efficient per unit area at reducing riverine nitrates than the most effective land-based nitrogen mitigation strategies, including cover crops and land retirement.[1] Nothing substitutes for natural processes. Sometimes we only have to learn to stay out of nature's way.

Macroalgae and microalgae offer distinct benefits over other methods that have been proposed to restore the Earth's carbon balance, but nearly all the studies conducted on carbon capture from biomass energy to date have focused on only terrestrial sources of bioenergy — that is, land-based photosynthesis. Few have looked at the potential for carbon capture from marine photosynthesis.[2] Since the production rates for water plants are typically an order of magnitude higher than those of the most productive terrestrial plants, they can produce an equivalent amount of bioenergy or food in less than one-tenth of the area.[3]

Could the warm-water coasts of Australia, Brazil, India, Mexico, the Middle East, Saharan North Africa, and southern Africa become the epicenters for clean energy and carbon rebalancing of the twenty-first century?[4] Although seaweed biochar has a low-carbon content compared with biochar from lignocellulosic materials, it has higher concentrations of macronutrients (nitrogen and phosphorus) and micronutrients (calcium, magnesium, potassium, and molybdenum) that are essential for plant growth.[5]

Farm trials of seaweed biochar in Australia produced results that astonished researchers. The average growth rate for treated test plots was 60 percent higher than the average growth rate for the nontreated plots, translating into an average of fourteen times more biomass accumulated over the course of the growing period. Their research paper concluded that algal biochar "could provide a significant revenue stream as a soil ameliorant and fertilizer, beyond its direct value as a tool for water remediation, and long-term soil C sequestration."[6]

Scientists at Cornell University have calculated that the current US liquid fuel demand could be met by growing microalgae in an area just over half of the size of Texas ($\sim$392,000 km^2/151,352 mi^2), and the current global liquid fuel demand could be met by an area of ocean slightly less than three times that size ($\sim$1.9 million km^2/733,594 mi^2).[7] The potential for co-production of protein and nutribiotics for animal and aqua feeds as well as direct human consumption is also substantial. According to a recent analysis,[8]

From the same 1.92 million km^2 needed to meet the current global liquid fuel demand, 2.4 gigatons of protein can be co-produced. This corresponds to about 10 times the total annual global production of soy protein. In addition to the potential significance of these nutritional coproducts to global food security, their high value will enable microalgae biofuels to become cost competitive with fossil fuels. Even using the current base-case, dry biomass productivity of 23 g/m^2 /day, the coproduction of aqua feeds can bring the cost of biocrude down to below the U.S. Department of Energy's near-term research target of $5 per gallon (3.8 L) gasoline equivalent (GGE). Target scenarios that bring this cost down to below $3 per GGE are anticipated for mature technologies by 2022.

As the authors of the study note, "Current agricultural demands for phosphorus are unsustainable, and global food security is already at risk this century unless society can become much more efficient in its use of fertilizers and recycling of nutrients from wastewater. Fortunately, the cultivation of marine microalgae can be highly efficient in its use of nutrients, only losing those that are actually harvested in the desired products."[9] The harvest need not be lost, either, if it is fed to animals, fish, or people and eventually returned to the soil as biofertilizer.

An even stronger case for co-production of nutrient-dense, macro- and microalgae animal and aqua feeds can be made due to the avoided CO_2 and N_2O emissions from conventional agriculture and the arable land freed up for production of crops destined to be consumed by humans, not animals.[10] While algal-based biopetroleum could be how soccer moms move around in the twenty-first century, and also how we replace plastics and polymer construction materials with non-fossil alternatives,[11] food security, land restoration, and other sustainable development goals may ultimately lead us to regeneratively farm the world's oceans, where life began.

The food, energy, and climate mitigation potential of seaweed is enormous. In one small 9.5 km (5.9 mi) stretch of Kenyan coastline, 6.8 kilotons of seagrass (dry weight) has been harvested yearly.[12] Pyrolyzed, it yields 48 to 57 percent biochar, higher than many other types of biochars. Using more conservative ranges (a 25 percent biochar yield is far more common,

coveralls with hood, hooded tunics, mitts, masks, and face shields with thermal suppression of the eyes and at least a portion of a wearer's face, as well as a breath-suppression device.

In Japan and Korea, mattress and pillow mats lined with bamboo charcoal absorb toxins (such as formaldehyde from glues used in furniture, automobile exhaust, or city air pollutants), bacteria, body odor, and electromagnetic wave radiation. They have antifungal properties, trap pollen and mold, and reduce the incidence of dust mites that can trigger allergies and asthma. Charcoal's ability to keep humidity levels low is lethal to molds and mites. The products are also said to help those suffering insomnia, headaches, fatigue, rheumatism, and poor blood circulation.[9] A few hours in the sun every few weeks refreshes the charcoal within the mats.

When biochar was rediscovered in the early part of this century, much of the excitement focused on carbon sequestration. Soil scientists highlighted the cobenefits of increased soil fertility and safe, long-term carbon drawdown. If you were to look at where the biochar industry is today in some countries, it is less focused on improving soil carbon or rebalancing greenhouse gases, and more focused on reducing air pollution and management of stormwater, manures, sewage sludge, and crop residues.

Next-Generation Paints and Plasters

When organic winemaker Hans-Peter Schmidt began experimenting with biochar in 2008, he intuitively understood that it would help build the *terroir* of his Mythopia vineyards. Schmidt is a biochar pioneer who has been making biochar, biochar kilns, and biochar composites and using them in real-world scenarios longer than most. Although his initial focus was on testing biochar in soils, his curious mind would not be satisfied. Pretty soon, the rooms of his four-hundred-year-old house in the Swiss mountains were sporting fresh coats of plaster infused with biochar.

In combination with the right ventilation, biochar plaster is able to keep cellar humidity at a constant 60–80 percent throughout the year, preventing, or at least greatly reducing, the development of unfavorable molds and odors. At higher humidity levels, the walls quickly adsorb moisture, returning it to the room just as quickly when humidity levels drop. Given that high humidity promotes the development of harmful microbial environments and low humidity is also not beneficial and can lead to particulate pollution, electro-static charging of the air, and the evaporation of wine in their wooden barrels, the ability to respire humidity should be highly desirable to vintners. Optimal humidity promotes healthy microbial flora in the cellar, enhancing the wine.

The result of Schmidt's first experiment was a gray-black plaster with a slightly shimmering appearance. To get a lighter color, Schmidt discovered that the portion of finely ground biochar could be reduced for the top layer, or any clay-based paint could be applied as the final coating. The following year he repeated the process at the Ithaka Institute — an international research foundation dedicated to carbon strategies and biochar in particular — this time taking scientific measurements of indoor flora and climate. He could verify in fine detail that biochar in the clay plaster helped adsorb contaminants, spores, and mycotoxins, as well as modulate gases.

In "The 55 Uses of Biochar," which Schmidt published first in his own *Ithaka Journal* and later in ten other publications,[10] he and coauthor Kelpie Wilson listed several benefits of biochar-amended plasters and paints:

- Insulation
- Air decontamination
- Noise reduction
- Low electrostatic charging of air
- Conservation of wood
- Reduction of dust and dust mites

- Antibacterial
- Flame retardant
- Restoration after floods
- Humidity regulation
- Odor reduction
- Electromagnetic radiation ("electrosmog") shielding

On the other side of the world, the Japan Housing and Wood Technology Center discovered that biochar has traditionally been used in Asia for wood-framed houses built on elevated foundations. Laid over the surface of the crawl space below ground floor, biochar controls moisture, termites, and mold and ensures that wood floors and supporting beams remain healthy and intact longer.[11] In Japan, people also use clay-biochar renders, the first coat of plaster, over and around radiant heat conduction pipes in walls to help retain and evenly distribute temperature. Because there is no access to oxygen or ignition sources within the wall, and because the biochar is buffered with at least 50 percent clay and sand, there is no risk of catching fire when these pipes supply steam heat.

Schmidt has inspired others (including us) to start experimenting with biochar plasters. In mid-2017 a small crew of eager learners attended our "Biochar from the Ground Up" workshop at The Farm in Tennessee. They helped mix biochar, sand, and lime and then slather it onto the bottle walls and bancos of timber, cob, earthbag, and strawbale buildings known as The Prancing Poet and The Green Dragon.

The key quality when adding any coloring agent to paint, stain, or spray plaster is it must be finely ground. Our first batch used coffee chaff char, which turned out to look a bit like black stucco or perhaps asphalt. It had a lot more surface area, and that led us to another discovery. Biochar paint will absorb heat very efficiently. We tested Kathleen's finer-grained biochar made from grape seed husks, and that produced a much darker shade of black paint without the texture. (While doing that Kathleen happened to have poison ivy for the second time in a month, so she decided to paint some on the affected area. It stopped the itch.) This paint, whether fine or tarry, might do well as a post-pruning or wound care covering for tree limbs.

Lest you think putting biochar in plasters and paint will only yield fifty shades of gray and black, this is not the case. In experimenting with clay-based paints, Albert mixed different proportions of biochar powder and got different colors. A 5 percent biochar, 5 percent sand, 40 percent clay, and 50 percent aged lime putty mixture created a dark green wall. The various tints of natural clay lend a range of hues to finished walls and furnishings.

Schmidt is enthusiastic about the potential. Biochar paints and plasters offer great potential for a variety of enclosed spaces, including food stores, animal housing, warehouses, or even houses and offices. It can clean up "sick building syndrome." The filter effect can absorb pollen, mold spores, dust, and cigarette smoke.

Optimal humidity levels have a major health impact inside houses and offices. A humidity level below 40 percent can lead to dry mucous membranes, increasing the risk of colds, asthma, and allergies. Conversely, a humidity level above 70 percent in closed living spaces increases exposure to mold spores. Toxic mold may cause depression, kidney and liver failure, decreased brain function, heart disease, eye irritation, headaches, vomiting, impaired immune system function, and severe respiratory distress.

Homes that have flooded, or even those with small leaks under a subfloor or in the walls, can create an environment where mold can thrive and mold spores circulate from room to room. Bathrooms, basements, and laundry rooms are particularly prone to mold.

As we learned from Hans-Peter Schmidt's wine cellar, biochar can eliminate mold growth simply by being added to paints and plasters. Biochar is one of the best ways we know to turn a sick building back into a healthy one.

David Derbowka, founder and chief executive officer of Passive Remediation Systems in British Columbia, is an expert in rhizoremediation (mushroom-based soil decontamination) and phytoremediation (plant-based soil decontamination) of groundwater, sludges, soil vapors, and sediments. For the past ten years Derbowka has been remediating landfill leachate with hybrid poplar trees. But what do you do with ten-year-old poplars that may contain toxic residues?

Pyrolysis provided an intriguing solution. The pyrolysis process purged the poplars of the harmful chemicals and immobilized any metals absorbed during phytoremediation. Still, even if not bioavailable, the heavy metals in the biochar would very likely prevent it from being land-applied, as there are restrictions on the amount of contaminants allowed in soil amendments.

Derbowka was determined to find a useful purpose for his char, so he started making various prototypes of drywall using biochar and plaster and bricks made solely with biochar and cement — saving the sand that he sees being harvested from local rivers. Testing it for humidity control, electromagnetic shielding, fire resistance, and other relevant properties is ongoing.

CHAPTER 16

Carbon Detox

There are biochar aficionados and then there are chardcore charistas. Kathleen used to think she was in the latter category until a friend informed her he consumes biochar with yogurt on a daily basis. Then another friend told her that she is a char chewer and has started to brush her teeth using biochar. Although we were aware of many of the benefits of feeding biochar to animals and knew that vitamin shops sell activated carbon for detoxing, for some reason dining on carbon-infused foods and drinks hadn't come up on our radar. Could this one minor change to personal habits be one small step for mankind and one giant leap to reversing climate change?

Human health and well-being are facing unprecedented threats from our collective mismanagement of inorganic substances like heavy metals, radionuclides, and antibiotics. Add to that the sterilization of our microbiomes and the in-migration of pests and pathogens from exhausted soils and denuded woodlands, or the antibiotic-resistant "superbugs" appearing in hospitals and blamed on livestock dietary mismanagement. Populist deregulation and regulatory capture by Big Pharma, the largest lobbying force on the planet, have elevated these dangers. Carbon can play an important role in mitigating the damaging impact of toxins and pathogens and also cut air and water pollution. By going straight to the source, cascaded carbon can displace the very products that are causing harm inside humans.

The anime TV show *Yakitate!!* is about a child's quest to create a national bread for Japan. The twenty-ninth episode is about bamboo biochar bread. An evil bakery rips off the diet-bread recipe of the good guys, who then counter with a bamboo charcoal bread. Spoiler Alert!: The bad guys are left in tears at its flavorful beauty. Somewhere in there they also manage a giant robot battle. We've found charcoal in Japanese desserts, real-life charcoal bread, bamboo-biochar-crusted confections sold in Japanese vending machines, and a restaurant in Vancouver that has bamboo charcoal ramen.

One time Kathleen was tabling at an event in the Finger Lakes region in New York State when a woman approached and began gushing about biochar. Kathleen asked why she was so enthusiastic. She said that she'd been suffering from mercury poisoning for a long time, had tried many different treatments to no avail. Finally, a holistic practitioner asked if she'd be willing to try biochar as a detox option. Having been told there were few possible side effects, she was more than willing to give it a go. The therapy was small doses for three days, then wait three weeks and take another round for three days. Blood work was conducted before and after. Not surprisingly to us, the mercury was gone. She was happy and finally healthy.

Kathleen asked if she meant activated charcoal or actually biochar. She said, yes, it was definitely biochar. So, where had she bought it? She gave the name of a local woman that Kathleen had sold a small amount to the year before to test on a few dairy cows that were not faring too well. That batch had been made from grape seed cake, a bit hard to come by unless you happen to know someone pressing grape seed oil. Kathleen suspected that the batch of biochar that detoxed the woman with mercury poisoning was the same intended for the bedridden bovines.

A cautionary note to anyone taking over-the-counter or prescription medicines: Ingesting activated carbon or biochar too soon after medicines can neutralize their impact. Ingesting it on a daily basis may also contribute to backing up the pipes, so to speak. We recommend speaking to a medical professional before commencing any charcoal-based detox regime. We would also not advise selling homemade biochar to someone for human consumption without further testing. And you should not consider using biochar, or activated carbon, when petroleum, alcohol, lye, acids, or other corrosive poisons are ingested. Carbon doesn't absorb these toxins like a sponge. Instead, it works through the chemical process of adsorption, in which elements bind to its surface.

Still, we know activated carbon is effective at absorbing mercury and is used in hospitals for that purpose and for many other poisonings and overdoses. Since biochar is basically in the same carbon family and generally less expensive to produce than activated carbon, focusing on the many uses for activated carbon that biochar could substitute for, from remediation to water filtration, is a logical market for lower-cost, easy-to-make biochar. The porous surface of biochar has a negative electric charge that causes positive-charged toxins and gas to bond with it. The nooks and crannies adsorb the toxins, allowing them to be discharged from the body along with the biochar. Most organic compounds, pesticides, mercury, fertilizers, and bleaches bind to biochar's surface, allowing for quicker elimination, while preventing the absorption in the body.

Activated charcoal is also used in the event of an accidental, or purposeful, overdose of many pharmaceutical drugs and over-the-counter medications. It's effective for aspirin, opioids, cocaine, morphine, and acetaminophen. It's important that the proper amount is administered as quickly as possible — within an hour of ingestion. According to the University of Michigan Health System, 50–100 grams (*not milligrams!*) (1.76–3.5 ounces) is used in cases of poisoning in adults and 10–25 grams (0.3–0.9 ounce) for children. Activated charcoal can also be used in cases of food poisoning when nausea and diarrhea are present. Adults take 25 grams (0.9 ounce) at onset of symptoms or when food poisoning is suspected, and children should be given 10 grams (0.3 ounce). Increase dosage as necessary. It's essential to drink an adequate amount of water when using activated carbon.

While activated carbon does not adsorb alcohol, it does help quickly remove other toxins from the body that contribute to poisoning. Alcohol is rarely consumed in its pure form; mixers that include artificial sweeteners and chemicals are common. Activated charcoal removes these toxins, too. In addition, when activated carbon is taken simultaneously with alcohol, some studies show it can significantly reduce blood alcohol concentrations. Princeton University's *First Aider's Guide to Alcohol* indicates that activated carbon is administered in some situations related to alcohol. This includes nasogastric intubation if the individual is unconscious or showing signs of acute alcohol poisoning.

What do you do with drugs when they are past their expiration date or otherwise in need of disposal? Unused or expired drugs, in liquid or pill form, can be poured or dissolved into biochar and the drugs will be absorbed into the pores. It is still advisable to keep that drug-saturated char out of the environment, but containing the drug in biochar is still an improvement over flushing biologically active liquids down drains or sending them out with the garbage.

Diatom Delicacies

In livestock feed biochar could replace diatomaceous earth, added as a binding agent for toxins. Diatomaceous earth is the hundred-million-year-old remains of marine phytoplankton. It can be found in toothpaste, anticaking agents, cat litter, termite barriers, and paint. It builds, absorbs, insulates, cleans, fills, feeds, dilutes, combusts, filters, and grows. The qualities of diatomite were explored in the early nineteenth century, but it first came into large industrial use in 1867 when Alfred Nobel chose to use it as a stabilizer for the nitroglycerin in his invention, dynamite. This brought fame and glory to the chalklike, soft, friable, earthy, fine-grained, siliceous sedimentary rock. Many new uses were discovered. It is finely porous, low in density (floating on water at least until saturated), and essentially chemically inert in most liquids and gases.

There are more than ten thousand species of extinct and living diatoms, some of which live in ice or hot springs as well as marshes or even on moist bark, but they are most abundant in sunlit fresh water or salt water, constantly enriched with suitable nutrients and dissolved silica. Live cells are covered by a jelly, and masses appear as brownish water or films on stream bottoms. When they die, they form the rich sediment known as diatomite, or diatomaceous earth. This process is ongoing. Worldwide, there are estimated to be 800 million metric tons available for mining, or 364 times the current annual removal of 2.2 million metric tons.

Diatomaceous earth is used in water filtering, food manufacturing, skin products, and farming to naturally eliminate free radicals, viruses, parasites, and other harmful organisms by binding to them and drying them out. It is harmful to insects with exoskeletons such as ants, beetles, and wasps because it abrades their exterior. Those insects learn to avoid it. It also has the ability to improve the body's use of calcium, improve bone mineralization, protect joints, and fight effects of aging. And that's not all. It reduces odors, curbs gas, cleanses the digestive tract, boosts liver function, and absorbs harmful toxins within the blood. It slows oxidative damage and has antiaging effects.

High-grade silica is refined from diatomaceous earth to make solar panels. The silicates in diatomaceous earth have strong aluminum affinity and that helps eliminate heavy metals from the body, including aluminum, making it useful for a heavy metal detox. While carbon does not have the

Table 16.1. Carbon Math: Special Earths Displacement

Global production diatomaceous and fuller's earth	5.5 Mt/yr
Impact of mining	0.1 MtCO$_2$-e/yr
Replace 20% with biochar	1.1 Mt/yr
With 82% C content	0.9 MtC/yr
CO$_2$-e drawdown potential	(3.3 MtCO$_2$-e/yr)
CO$_2$-e avoided from mining	20 ktCO$_2$-e/yr

Note: Potentials for atmospheric removal are in parentheses. The degree of substitution is based upon the physical properties of the substance being displaced and the authors' estimate of market penetration potential.

same high aluminum affinity, it does bond with other heavy metals, including some that silicates might have a harder time with.

However, as organic as diatomaceous earth is claimed to be, it is still a mined product that brings with it the carbon footprint of mining, milling, and shipping: first to the manufacturer, then to the consumer, then to a landfill. Miners and millers have a significantly higher mortality from lung cancer than the general population.[1] Biochar has other advantages when it comes to environmental benefits and impacts. Since the largest uses for diatomaceous earth are filtration (50 percent), fillers (25 percent), insulation (17 percent), and absorption (10 percent), we see a market opportunity for cascaded carbon to displace this mined material. And while diatomaceous earth users, who consume 2.2 megatons per year, may seem like a large potential customer base, cascade designers might also look at fuller's earth, a soft abrasive used for grease removal and as polish for silver and chromium, with global production at a not-too-shabby rate of 3.32 million tons per year.[2]

The environmental impacts of diatomaceous and fuller's earth mining, milling, and transportation are large. Abandoned mining sites in the United States release 1.6 megatons of CO$_2$ annually. This is dwarfed by fugitive methane emissions that can be 70–95 percent of all emissions during operation and retirement combined. China releases 9,600–12,000 cubic meters of toxic gas from mining and milling for each ton of rare earth elements produced. Additionally, nearly 75 cubic meters of acidic wastewater and 1 ton of radioactive waste residue are generated *per ton*.

If we look at the climate impact of silver polish and solar panels, we come up with some pretty astonishing numbers. With 5.5 megatons of diatomaceous and fuller's earth produced each year, we can make an educated guess and suggest that mining those is not very different from similar kinds of mining, and therefore will generate about 100,000 tons of CO_2-equivalents annually. Replacing 20 percent of the market with biochar, which actively *removes* CO_2, would require 1.1 megatons of high-carbon biochar, which translates to a *removal of* 3.3 megatons of CO_2-e.

Air Freshener

Most quartz sand is excavated for glass and concrete, but no less significant are hundreds of other uses, from computer chips to bathroom tile. Once the sand has been sifted, washed, scrubbed, acid-leached, sifted again, and transported, it begins to undergo its many transformations. It changes into, among other things, silica gel, a desiccant used to control moisture and mold.

Thousands of tons of desiccants are manufactured every year and thrown away after a single use. They are packed into the shipping boxes that bring you products from Amazon and Alibaba. Desiccants are used to dry-store everything from bowling shoes to military munitions. Open a pill bottle, a boxed electronic device, or a new gym bag and what do you find? A tiny little sachet of desiccant, there to protect the contents from mold and odors. Open the porous packet and what's inside? Silica beads.

Why silica? When pure sand, sodium silicate, is treated with sulfuric acid, the resulting beads resemble biochar — spherical asteroids of pitted pores with an outsized surface area that will attract and hold water by adsorption and capillary condensation. Silica gel, like biochar, will adsorb about 40 percent of its weight in water. Like biochar, it is noncorrosive and nontoxic. Yet unlike biochar (but like clay, another common desiccant), silica will begin to lose its adsorption capacity when temperatures rise above 77°F (25°C).

It is unlikely that desiccants would be such a huge market today — $2 billion worldwide and doubling every ten years — were it not for some clever branding that changed "desiccant" into "air freshener" in 1948. We spoke with Eric Rubin, creator of the all-natural Moso Natural line of air fresheners with bamboo biochar as the sole ingredient. He told us he had grown up working in his father's automotive parts store and couldn't help but notice how many air fresheners they sold. "Air fresheners own the market," he told us.

When he looked inside the air fresheners, however, he found a lot more than plastic beads. The basic ingredients are formaldehyde, petroleum distillates, 1,4-dichlorobenzene, and aerosol propellants. The US National Institutes of Health found that regular exposure to 1,4-dichlorobenzene is harmful to the lungs.[3] Formaldehyde is a carcinogen.[4] Aerosol propellants can damage the ozone layer.[5] And then there are the fragrances. A study performed by the Natural Resources Defense Council (NRDC) found twelve of fourteen common household air fresheners contain phthalates, although none listed phthalates on their label.[6] At least in the United States, air freshener manufacturers are not required to list ingredients.

According to the NRDC:[7]

- Di-ethyl phthalate (DEP) is associated with changes in hormone levels and genital development in humans.
- Di-n-butyl phthalate (DBP) is recognized as a reproductive toxicant by the National Toxicology Program and the State of California. It can lead to changes in genital development.
- Di-isobutyl phthalate (DIBP) is associated with changes in male genital development.
- Di-methyl phthalate (DMP) has been linked to reproductive toxicity in animal studies.
- Di-isohexyl phthalate (DIHP) is likely a developmental and reproductive toxicant, according to limited toxicity testing.

Professor Anne C. Steinemann at the University of Washington investigated top-selling air fresheners and laundry products and learned, as NRDC had, that air fresheners contained more than twenty different volatile organic compounds, with more than one-third classified as toxic or hazardous under federal laws, including carcinogens with no safe exposure level.[8] None of these chemicals were listed on any of the product labels or Material Safety Data Sheets. Chemicals included acetone, the active ingredient in paint thinner and nail-polish remover; acetaldehyde and 1,4-dioxane, both carcinogens; and chloromethane, a neurotoxicant and respiratory toxin. Even air fresheners labeled "organic," "green," or with "essential oils" emitted hazardous chemicals, including carcinogens. In 2009, Steinemann teamed with Stanley M. Caress of the University of West Georgia to conduct epidemiological studies. They found nearly 20 percent of the general population and 34 percent of asthmatics report headaches, breathing difficulties, or other health problems when exposed to air fresheners or deodorizers.[9]

Rubin told us that when he first started introducing his all-natural biochar alternative, he faced tough competition. "I did my first trade show in August 2010. I got twenty-three stores. Then I did another show in November, got another twenty stores. January 2011 was rough, I only did about $2,000 in sales."

His father doubted the venture would succeed, but his wife-to-be encouraged him. "By August I was doing $20,000, and that's when I really started picking up steam. I did fifteen trade shows that year. Twenty twelve was my first successful full year.

"Natural odor eliminators are far from the norm," he told us. "This will change in the future, but it will take fifteen to twenty years. The fragrance business is powerful. It isn't until people understand what air fresheners are really doing to them that the market will begin to accept natural alternatives. Think of [the switch from] Coke to bottled water."

Today Moso Natural is a thriving business. Eric Rubin's father is joining him as a partner in a new product line. The future for natural air fresheners is bright.

The desiccant market goes well beyond air fresheners. Dessicants are applied to newborn farm animals, such as piglets, to dry them quickly and prevent infections, which can be crucial for the animal's survival. They reduce bacteria and pathogens that thrive on wet surfaces. They also make it more difficult for mold spores to mature into molds. Farm products like cocoa, coffee, and various nuts and grains are particularly susceptible to mold and rot when exposed to condensation and humidity. Often they are wrapped in corn husks or paper in nonindustrial countries, or packed in some form of foam absorbent in the industrial world, to protect them from mold and bruising while they travel. Wouldn't it be better if farmers who need to ship fruits or vegetables long distances simply made their own desiccants just by carbonizing their unused agricultural waste?

Table 16.2. Carbon Math: Clay Displacement

Global production of bentonite, kaolin, and palygorskite clays	50.6 Mt/yr
Portion devoted to air fresheners	1.0 Mt/yr
Replace 20% with biochar	0.2 Mt/yr
With 82% C content	0.16 MtC/yr
CO_2-e drawdown potential	(0.58 $MtCO_2$-e/yr)
CO_2-e avoided from mining and transportation	unknown

Note: Potentials for atmospheric removal are in parentheses. The degree of substitution is based upon the physical properties of the substance being displaced and the authors' estimate of market penetration potential.

If shoe or electronics manufacturers begin to consider the life-cycle impacts of their industry, couldn't they reasonably change from silica gel packing material, which generates greenhouse pollution, to biochar packing material? Biochar can also easily substitute for most of the other major desiccants in use today: bentonite or montmorillonite clay; zeolite (also known as molecular sieve or aluminosilicate); calcium oxide (calcined or recalcinated lime); and calcium sulfate (known commercially as Drierite, created by the controlled dehydration of gypsum). Unlike some desiccants, biochar may even offer thermal insulation properties.

Desiccants are the part of air-conditioning systems that dry the air to improve and preserve the efficacy of the refrigerant and reduce corrosion from condensation. The whole new industry of energy-efficient window glazing relies on zeolite spheroids between multiple panes of glass to prevent condensation. In a business landscape dominated by mined clays and fossil-carbon plastics are emerging opportunities to eliminate toxins, sequester carbon, reuse, and recycle. As with so many other biochar products, once dehumidifying duties are done, greater and longer cascades await, until finally, its last job done, it goes back to the ground as a carbon sink.

PART IV

Carbon Conversion: Cascades in Action

Spring passes and one remembers one's innocence. Summer passes and one remembers one's exuberance. Autumn passes and one remembers one's reverence. Winter passes and one remembers one's perseverance.

— YOKO ONO

CHAPTER 17

Cryptocarbon

I f, after reading this book, you can envision how climate might be tamed within your lifetime, you can see how the technology pieces together and how it can pay for itself from the business returns it creates, then what is the impediment? Answer: inertia. Specifically, social and cultural inertia. Ultimately the solutions to the climate crisis do not depend upon technology. They come down to people making reasonable choices. The big question is whether people will adopt the changes we have outlined in this book.

Governments come together in meetings and draw up "climate finance" plans to fund carbon drawdown projects. For-profit and nonprofit initiatives spring up along the same lines, and we feel comfortable in predicting their pace will quicken as supernormal climate events compound the urgency. The United Nations calculates that the cumulative financial need of developing countries to mitigate and adapt to climate change is approximately $5.7 trillion,[1] mainly from Asia (50 percent) and sub-Saharan Africa (42 percent).

The $100 billion per year pledged from developed to developing countries at the Copenhagen Summit in 2009 and reiterated at Paris in 2015 is woefully insufficient. That level of funding would likely guarantee a temperature increase, relative to preindustrial levels, of about 3.2°C (6.8°F) by 2100. The world would need to double its annual investment over the next fifteen years just to hold to two degrees under the best-case scenarios for mitigation potentials. At least a third of that needs to be directed toward decarbonizing infrastructure.[2]

As we have traveled around gathering material for this book we have seen scores of "shovel-ready" projects, from rural cooperatives to urban ecovillages to international ecosystem restoration camps for engaged youth. These programs are all good candidates for acceleration by the Green Climate Fund, Climate CoLab, Common Earth, crowdfunding, and other financing mechanisms. Actual working examples are vital. They demonstrate what is possible and cloneable. But we began to wonder if the old model of donations and loans is not itself part of the problem.

Whether you are a winner or a loser in the current world monetary system, you have to admit it is not serving the planet very well. An ecological economist would say the planet is suffering because there are too many "neglected externalities," such as whales, pollution, the cost of nonrenewables, and the real needs of people for health and happiness; all are subjugated to the desire to maximize profit.

As long as the international fractional reserve, "fiat currency" banking system — really just a Ponzi scheme with vastly more claims on the underlying real wealth (natural capital) than could ever be serviced — remains the only option, the future looks bleak. Not all economic systems in history have been this messed up, but for as long as some type of debt jubilee or new, greener economics has been called for, we have remained mired in the *status quo ante*.

Until now.

Say hello to the blockchain. The number of initial coin offerings will make 2018 "the year of the cryptocurrencies." Until 2019 exceeds it.

But what does this mean for a more sustainable, steady-state, circular economy, one that meets goals of health and happiness? Aren't blockchain-based tokens just a way to dupe gullible speculators into investing millions into nothing more than a white paper, a website, and an algorithm? Equally troubling is the inordinate energy consumption of Bitcoin. The Bitcoin network is estimated to have consumed at least 2.55 gigawatts of electricity at the start of 2018, and will take potentially 7.67 gigawatts by the end of the year, placing it somewhere between Ireland (3.1 gigawatts) and Austria (8.2 gigawatts). That's roughly 0.1–0.3 percent of global generation, or one hundred times the amount of renewable power generation installed in North America in the past decade.

The primary culprits are Bitcoin mining appliances like the Antminer S9, a computer processor that does nothing but endlessly crunch algorithms to lengthen the blockchain. When Satoshi Nakamoto presented his solution[3] for securing digital transactions in 2009, he assumed that its inflation would be limited by the cost of electricity. Nakamoto's solution involved "timestamping transactions by hashing them into an ongoing chain of hash-based proof-of-work." Bitcoin miners receive coins as payment for making hash calculations. As long as the coin value exceeds the price of electricity, there is an incentive for anyone to head for the nearest Walmart Superstore and plunk down $8,200 for an Antminer.

As of mid-March 2018, about twenty-six quintillion hashing operations are performed every second and nonstop by the Bitcoin network. The primary fuel for each of these calculations is electricity. A single Antminer draws a load of 1.5 kilowatts — enough to power two refrigerators and a flat-screen TV. If you run an Antminer 24/7 for a year, it will require about 15,000 kilowatt hours — from $450 (at 3 cents per kWh) to $1,800 (at 12 cents per kWh) to mine one coin. Bitcoin was trading at under $1,500 until May, 2017, when it began a rise that took it to over $19,000 by December. It fell back to a trading range between $6,000 and $12,000 in 2018, but there's still plenty of profit to be made for miners.

Cheap electricity has lured miners to set up data centers near hydropower facilities in secluded regions of China, Canada, and Norway or beside geothermal wells on volcanoes in Iceland. The electricity demand for Icelandic Bitcoin mining — around 840 GWe — has already surpassed the total combined electricity use of every Icelandic home — 700 GWe.[4]

And let's face it: Outside of Iceland, Bitcoin miners, for all their geeky pride, are really coal miners in disguise. But they are not the only ones. By 2040, if not sooner, electric cars could require more than 280 gigawatts of additional power capacity. Translating into solar panels, that would cover an area of 55–148 square miles (142–383 km²). Or it would take 28,000 to 56,000 of the largest offshore wind generators being manufactured today.

Number seven on the UN's list of Sustainable Development Goals is affordable and clean energy. Putting the internet in everyone's pocket means more than the demand for recharging power will rise — aspirations will also rise. The more power people have, the more people will demand. Fossil fuel electric generators must be retired, as quickly as they can be replaced. At the moment, even though new solar power has dropped to less than fossil operating cost, and far less than nuclear energy, with greater advantages for public health, grid reliability, baseload, and national security, solar farms are still more costly to build, and with a larger embodied fossil energy footprint, than biomass energy with carbon cascades.

What if, instead of exponentially increasing power demand, and atmospheric carbon, the next world monetary system exponentially decreased it? How about if, at the same time, it worked to achieve all the sustainable development goals and brought population and sprawl under control so that a better world could actually be *sustained*?

Currently, carbon-removal approaches, like renewable-energy strategies, have different costs, degrees of maturity, permanence, and uncertainties. The blockchain — really just a verification and tracking ledger — could allow us to establish, harmonize, and stabilize a market approach to these innovations. Algorithms could be written to include social and ecological goals in a blockchain's DNA. Maybe they could even be designed in such a way that the verification process would not consume ridiculous amounts of energy, such as by using proof-of-stake rather than proof-of-work (experimenters call this Bitcoin Green[5]). One coin in development is the Nori, worth 1 ton of CO_2 removal. Its creators plan to use it to restore the atmosphere. At the Collision! conference in New Orleans in May 2018, Albert spoke with Nori Chief Development Officer Christophe Jospe about the coin.

"Whenever a seller performs a carbon-removal action, we first verify the carbon dioxide has actually been removed from the atmosphere and stored," Jospe said. "The seller either uploads data manually, or the removal data is automatically reported into the Nori platform. This generates the creation of a new carbon-removal credit smart contract now available for purchase by a buyer. We only accept carbon-removal projects. Reductions in emissions are not good enough."

Another solution has been proposed in a September 2018 white paper by the Regen Network.[6] Regen applies "consensus power" using a platform called Tendermint that recruits a set of private validators, each of whom is known to all other validators and constrained by them. Otherwise known as a Proof-of-Stake consortium, this is inherently more energy efficient and faster than public Proof-of-Work systems. Validators earn fees for providing services to the Regen Ledger. The set of validator nodes (computers) in the consortium is constrained to be equal to or smaller than the number of consortium members to limit energy consumption. Every validator node must also stake a certain amount of tokens in a bond that can be forfeited partially or entirely in the case a validator fails in their responsibility to uphold network health.

The Regen Network plans "to create a global ecological accounting system and unlock the potential to reward increases in ecosystem health [by] creating a coordinated response to climate instability, soil loss, and biodiversity decline." The Regen Ledger would use three core review frameworks: Ecological State Protocols to verify a certain state or change of state on a piece of land; Ecological Contracts to fund and reward desired change in ecological state; and Supply Protocols to tie ecological state into supply chains in trusted ways that value local stewardship and community resilience.

Creating new cryptocurrencies is not a heavy lift. The hard part seems to be incentivizing the type of economic activity that leads to the kind of world we want to live in. Art Brock, one of the founders of the Holo project that attempts to slash the power demand of computing by allowing users to share spare capacity and be paid in tokens, says, "Cryptocurrencies do not have to be gambling tokens created from nothing. They can be responsibly connected to assets, promises, or real-world value. They don't have to re-create all the speculative money problems that they were supposed to be solving."

Going back to our earlier examples of asphalt highways, suppose a highway that was constructed with biochar substituting for aggregate was backed by a blockchain carbon credit. Decades later, that highway needs to be torn up and the question is whether to recycle the recovered aggregate or burn it to produce power. Burning it would transform it from credit to debit against the value of the country or the ownership entity. Recycling the asphalt would leave it on the ledger as an asset.

One proposal, from Hélène Nivoix in France to the UN Environmental Programme, is to set up a division called the Organic Monetary Fund. OMF would do something ecologist John D. Liu told us was the *sine qua non* for any serious drawdown attempt. It would peg to the growth of healthy living biomass, soils, and carbon underground.

All International Monetary Fund (IMF) member countries can volunteer to be part of the OMF scheme, provided that their country agrees to certified management cooperatives. These would be groups of small, multipurpose agricultural and forestry units applying the principles of agroecology and permaculture. The global scientific community provides annual figures on the growth of healthy living biomass produced by participating countries. The OMF then grants each member country a draw from the IMF in the relevant national currency. Under OMF rules, a nation's currency grows or shrinks in proportion to its role in the carbon cycle. Money is created by making land increasingly fertile. A nation's ability to borrow and lend diminishes if soil fertility is depleted and carbon is lost to the atmosphere or ocean. Every national economy is determined by how well each nation harnesses photosynthesis. Growth of soil fertility, biomass, and biodiversity means true wealth and abundance for local communities.

If you want to unravel the mystery of how we can take the great leap into reversing the course of climate change, just follow the money. In the near future, currency exchange might seem like a quaint notion, replaced by instant blockchain transactions that place value where it really belongs.

Engineered Disruption

Clayton Magleby Christensen's 1997 book, *The Innovator's Dilemma*, coined the term *disruptive technology* to describe the "process by which a product or service transforms an existing market by introducing simplicity, convenience, accessibility, and affordability."[7] Our personal favorite disruptive device is the smartphone. Not only did it disrupt the quaint landline and the legacy telecom industry, but it continues displacing other products: watches, cameras, boarding passes, toothbrushes, heart monitors, weather gauges, credit cards, and GPS navigators.

We weren't convinced biochar could be a disruptive technology until we understood how it could be blended with concrete, steel, wood, adobe, ceramics, and glass; until we understood that biochar can displace many "dirty" and sometimes expensive additives; and until we understood that sequestering carbon was not the only nor even, in many cases, the primary reason to include biochar in different composites or products. In the context of the need for humanity to make monumental shifts toward low-carbon-footprint materials and processes, when suddenly renewables and drawdown tools must displace anything still sending carbon skyward, biochar and carbonization begin to look uniquely promising, with massive disruption potential.

Biochar as a soil amendment is not particularly disruptive — it is even conceivable that Big Ag suppliers will simply add biochar soil amendments to their current chemical offerings to retain control of the fertilizer market and congratulate themselves for being greener. However, biochar replacing nonrenewable materials or materials that have a large carbon footprint could be hugely disruptive.

To gain purchase in the marketplace, disruptive innovations must first gain credibility. In the context of Aristotle's rhetorical triangle (ethos, logos, and pathos), ethos refers to credibility. Jonathan Shapiro's book *Lawyers, Liars, and the Art of Storytelling: Using Stories to Advocate, Influence, and Persuade* has a great perspective on ethos as it relates to Aristotle's rhetorical triangle: *Ethos* is often used to describe beliefs or ideals that define a community. Credibility can be boosted in many ways. Clothes credentialize cops, concierges, and candy stripers. Certifications credentialize professions and products. Testimonials by respected authorities or celebs boost credibility for a certain fan base. Shapiro calls this "ethos by extension." He observes "the field of expertise almost doesn't matter."[8]

Who or what to trust is not a trivial or rhetorical question, nor is there a simple answer. Scientists are often pushed to publish results that will win them tenure or grant money. Businessmen may make extraordinary and often unproven claims about performance. That is no way to lay the foundation for a paradigm shift.

Professor Dirk Messner at the German Development Institute argues any serious change in the arrangement of human civilization in the natural world of necessity would involve a redrawing of the social contract. "It will also require fundamental shifts and realignments in global institutions, global governance processes, and global civil society networks as a social contract for sustainability in a highly interdependent world," he says.[9]

One way this trust can be built is by application of blockchain as an agency of verification. The advantages:

1. globally accessible
2. trustless (no need for parties to trust each other)
3. fully secure (intemperable, incorruptible, nonhackable)
4. highly scalable
5. transparent
6. transactional
7. automated

8. immutable
9. deterministic
10. consistent and accurate
11. programmable
12. unstoppable

The disadvantage, besides perhaps that twelfth item, is that such a system once installed will be difficult to uninstall. It is also vulnerable to disruption by power outages. Nonetheless, it offers a bridge to cross the gap in trust and emerge with a viable new global economic system, one that reverses direction of the human carbon cycle.

Carbon substitutions are not applicable to everything everywhere, but for the right type of product or production technology, cool carbon substitutes can be both profoundly disruptive and marketplace champions. There are many ways to position these new cascades, but understanding what is most important to meet a particular need is the key consideration.

What more can we ask of life for a designer, planner, engineer, agronomist, merchant, or craftsman than this: to get to play the game at such a high and ambitious level, with such tools and teammates, and so much hanging in the balance?

CHAPTER 18

The Cool Lab

Sometimes it falls upon a generation to be great.
You can be that generation.

— NELSON MANDELA, 2006

I n January 2009, Seattle biochar producer Art Donnelly was invited to Costa Rica's Santos region to show them how to build his Estufa Finca cookstoves. Respiratory disease among the migrant coffee-bean picker population was particularly severe; it was the leading cause of death for children under the age of ten. The Estufa Finca is a tabletop gasifier that produces biochar instead of smoke and ash. It emits 92 percent less particulate matter and 87 percent less carbon monoxide than the traditional open cooking fire.[1] During the coffee harvest of 2010–2011, Donnelly, a team of volunteers, and local residents built and installed thirty-two stoves in migrant worker housing in Santos. Donnelly and his team then went farther south and began a project teaching stovemaking to the Bribri people of Costa Rica's Talamanca region. The Bribri are the stewards of one of the most biodiverse areas of the Earth.

In 2014 we visited Costa Rica where we met the Estufa Finca project team. We met an eco-forester who chars much of the debris from his permaculture forest and applies it to fertilize a wide diversity of trees. He also makes biochar soap and toothpaste. We met an organic cacao producer who converts forest litter and cacao pods into biochar and applies it mixed with mulch to his forest plants. We even found jewelry made from biochar — *joyas de biocarbon*.

When you approach San José by air, it's hard not to appreciate the sheer volume and diversity of forest in Costa Rica. Meanwhile vast amounts of soil wash down the streets, rivers, and mountains with every heavy rain. For too long Costa Rica has been a bit "deluge-ional" about erosion. They plow the mud off streets the same way people in cold climates plow snow. Now, in no small measure due to clear-cutting, Costa Rica has been suffering substantial droughts during the annual dry season. Chemical agriculture has also become pervasive; Costa Rica uses more pesticides than any other country — 18.2 kilograms per hectare of cropland — three times more per area farmed than runner-up Colombia — and has become dependent on imported oil to run its farm machinery. This has taken a toll on the environment, especially water and soil quality, and on human and animal health.

Over one hundred years ago, Señor Cecilio Lindo started to farm the area between the Irazu and Turrialba volcanoes. The altitude at 3,000–4,500 feet (1,000–1,500 meters) was ideal for coffee and macadamias and the bottomlands suitable for sugar. In 1945, a group of Costa Rican investors acquired Lindo's farm and founded Hacienda Juan Viñas. They built upon Lindo's tradition of growing and processing coffee, sugarcane, and macadamia nuts of the highest quality. The fertile soils and perfect microclimate yielded organic coffee of excellent body, medium acidity, and great aroma.

Juan Viñas became a prime example of a large-scale, successful closed-loop biorefinery that combines economic development with social enrichment. The Hacienda processes its sugar, coffee, and macadamia nuts in its own mill. To generate heat for the distillation process (evaporating the water out of the sugar water) they cook the sugarcane waste (bagasse). Instead of burning the bagasse to ash, they have a process that creates biochar. And not just a little biochar. They churn out 200 tons of biochar per day during harvest season. The company also produces 850,000 kilowatts per hour of electricity in its hydroelectric plant.

Their sustainable development goals don't end at the mill gate. The company provided more than three hundred houses to its employees, as well as health care benefits and training programs. It donates forty-five hundred textbooks per year to local children. It also funds Aproca, a provincial soccer program that gives free training to boys and girls as long as they maintain acceptable grades at school.

All the food waste from the town is collected and cocomposted with the Hacienda's cane, nut, and coffee biochar. Ninety percent of the biofertilizer is used on their own fields and the rest is sold under the Hacienda Juan Viñas brand to nurseries in nearby towns. Already the biggest sugar mill in

Costa Rica, they have now become the first carbon drawdown sugar mill in the world. They want neither carbon credits nor kudos. They are doing it because it makes great economic and ecological sense for the two communities they serve: human and terrestrial.

Juan Viñas is in the earliest stages of a process Santiago Obarrio in the Dominican Republic calls the "Cool Lab," a microenterprise hub that marries sustainable development goals and good business practices. Terms like *profit* and *wealth* take on deeper meanings when they extend to treat everyone as family. To better understand the Cool Lab, we should take a trip forward ten or twenty years and imagine an impoverished district somewhere in the world — it could be Costa Rica — that reaches out to a team of designers for help.

Let us imagine: A coffee-growing village risks being carried away by mudslides that follow brush fires where the forest has been battered by hurricanes and then cut down to open the sky for coffee bushes. What things are scarce? In no particular order:

- food
- water
- cooking fuel
- secure shelter
- energy

- productive employment
- biodiversity
- soil
- birth control
- health care

What things are overabundant?

- mud
- deforestation
- rain
- hurricanes
- earthquakes

- unemployed people
- coffee
- resentment
- mosquitoes
- climate change

Let's see which of these things we can match up and cancel out. What we are about to describe is a carbon cascade.

The design team observes that the hillside needs to be planted with vegetation. It is especially important that the hilltops be forested. A keyline analysis will show where water wants to go when it rains, and how best it can be held high in the landscape and directed both to subsurface flows and to dam storage for the dry season. Alley cropping along the contours follows hand-cut swales, or machine-cut where financial capital substitutes for social capital.

The berms can be planted with successional understory (in this tropical example, pineapple, cassava, ginger, allspice, coffee, and medicinal herbs), mid-level canopy (banana, papaya, moringa, cacao, tree legumes of mimosa, cassia, and pea subfamilies), and eventual overstory (coconut, ramon, samwood, mahogany, cedar, breadfruit, bamboo, and peach palm). Between the alleys will be seeded supergrasses like kernza, sunn hemp, pennisitum, pearl millet hybrids, brassicas, amaranth, and others, as well as familiar food crops including maize, yam, and beans where soils and water supply are well suited.

As much as possible, the planting process can be accompanied by biofertilizers having a high percentage of finely pulverized biochar, activated indigenous microorganisms, some immediate food for those microbes (such as composted food wastes and manures), and minerals keyed to redress local soil deficiencies or acidity. If these biofertilizers are not immediately available for the first plantings, they can always be added later.

Water in pond storage on the hillsides is edge-planted with *Acoris*, a plant that inoculates the water with a mosquito-larva-destroying resin. As the plant matures, pools and dams progress from being mosquito generating to mosquito decimating. In the lowlands, water that overflows from catchments above is directed to *chinampas*, constructed wetlands composed of alternating islands and channels and rotating between aerobic (horizontal and vertical flow reedbeds) and anaerobic (settling lagoons) seeded with aquatic and semiaquatic plants (taro, Chinese water spinach, lotus, azola, rice) and freshwater fish (aquaculture). *Acoris* for mosquito control can also be planted here, but the fish do most of that work already, so the plant is only needed in mudflats and places the fish cannot go.

Within the first season, the hillside mud problem is eased, deforestation is halted, and food scarcity begins to be alleviated thanks to the fast-yielding varieties of annuals, perennials, and fish. Productive employment can expand this system as much as available land permits, even on relatively steep hillsides. Resentment diminishes. Hope emerges.

Employing Ostrom's formula (see sidebar: The Wisdom of the Commons), most resource appropriators participate in the decision-making process, decisions are monitored, violators are sanctioned, conflicts are mediated, and multiple, cooperating layers of nested enterprises are set in motion.

Within the village a regenerative, biological energy system arrives to replace the fossil fuel (diesel electric) grid-based source that previously had supplied electricity only intermittently, occasionally frying phone chargers and boom boxes. This system consists of a biomass furnace, running on the woody wastes from coppice (coffee, moringa, and cassava), coconut, rice or other shell crops, pelletized supergrasses (sunn hemp, leucaena, etc.), and other biomass after extraction of leaf protein.

The loading dock at the biorefinery receives raw materials harvested from the farms. Leaves of tropical legumes[2] are taken by conveyor and chopped into 2-centimeter pieces, soaked in 2 percent sodium metabisulfite, disintegrated in a hammer mill, and pressed in a single-screw press. The expressed juice is heated with steam (produced by the furnace) and protein coagulum collected, centrifuged, and pressed, then spread in a thin layer on glass plates and dried in an air-filtered, dehumidified room. It is then collected as a powder and containerized to be used or sold as a feed supplement.

At its most basic level, high-protein, high-quality leaf protein fractionation is simple. Production is geared to consumption by farm animals to reduce food safety, preservation, and storage concerns. Later improvements can produce dried leaf extracts for human consumption, but higher capital costs are incurred and clean-room protocols by workers become essential.

Following leaf protein extraction, the dried mash from the press is used as a feedstock for the gas retort, where it joins other dried agricultural wastes: coppice wood, prunings, bamboo thinnings, pallets, coconut coir, and nut and rice husks. All this is pyrolyzed, the heat captured to both run the leaf protein process and produce electricity (directly, with a Sterling or Minto heat engine, or indirectly, with an internal combustion, gas, steam, or diesel-fueled engine), and co-products (fractionated volatile gases, wood vinegar) are drawn off before the final product — high-quality biochar — is left.

The biochar is quenched (if destined for soil use, preferably with urine or snuffed with manure because that adds nutrients), pulverized, and charged (blended with microbe-rich aerobic compost) to make a potent

The Wisdom of the Commons

Elinor Ostrom, the only woman to win the Nobel Prize in economics, was an Indiana University political scientist who spent fifty years cataloging evidence against Garrett Hardin's theory of the "tragedy of the commons." Ostrom showed that people can and do spontaneously self-organize to sustainably manage resources. She looked at mountain villages in Switzerland and Japan, forests in Nepal, irrigation systems in California and Spain, fisheries in Maine and Indonesia, and the polycentricity of police functions in Indianapolis. She found humans seldom feel trapped or helpless amid diminishing supplies; rather, they innovate. Ostrom argued against any singular panacea for social or ecological problems[3] and while we realize this book might sometimes seem like it's purveying biochar as a universal solution to climate change, we wholeheartedly agree with Ostrom's focus on the multifaceted nature of human-within-ecosystem.

Ostrom identified eight "design principles" of stable local common pool resource management:[4]

1. Clearly defined (clear definition of the contents of the common pool resource and effective exclusion of external unentitled parties)
2. The appropriation and provision of common resources that are adapted to local conditions
3. Collective-choice arrangements that allow most resource appropriators to participate in the decision-making process
4. Effective monitoring by monitors who are part of or accountable to the appropriators
5. A scale of graduated sanctions for resource appropriators who violate community rules
6. Mechanisms of conflict resolution that are cheap and easily accessible

7. Self-determination of the community recognized by higher-level authorities
8. In the case of larger common-pool resources, organization in the form of multiple layers of nested enterprises, with small, local, common pool resources at the base level

In the second half of the twentieth century and into the twenty-first, these design rules have been put into motion in a deliberate way by thousands of communities and organizations. They were applied in the Quaker-style consensus meetings of the antiwar and antinuclear movements, intentional communities, Solidarność, the Bioregional congresses, the Global Ecovillage Network, Transition Towns, Occupy Wall Street, and many others. Techniques have been iteratively improved until they've coalesced into a grammar of open, transparent, egalitarian decision making.

"cool" biofertilizer. Alternatively, it is kept at food grade and sold as a dry product for use as a food supplement, animal feed probiotic, water filtration medium, or deodorizer. It could still become biofertilizer, after undergoing one or more of these other uses. At less than food grade it can be used as a litter amendment to reduce smells in animal enclosures, improve the fermentation of silage, or go into various natural building materials — paints, dyes, plasters, wallboard, and bricks.

Styrofoam clamshell food containers, which are ubiquitous from take-out restaurants and shops in the cities and often wind up just floating away on ocean currents, never to be destroyed, are collected and brought to the biorefinery. There they go into an acetone bath, and the dissolved liquid is blended with low-grade biochar and poured into molds to dry. The resulting hard resin is moldproof, waterproof, nondegradable, lightweight, and durable. Depending on the dies and molds, it can become a variety of products — extruded lumber, roofing tile, chalk, surfboards, fishing boats, life vests, doors, bicycles, and ice chests. If there is a surge in demand for a particular product — refrigerator deodorizers or animal feed supplements, for instance — or there is a surplus of some particular feedstock — bamboo knocked down by a storm — the biorefinery can shift its production pattern to take advantage immediately.

This system sequesters more carbon than it emits, so we call it "cool." By adding biochar, mineral-rich compost, and microorganisms to nutrient-poor or eroded soils, we can jump-start soil productivity and boost farm incomes. The gains in those alley-cropped contours might be anywhere from 30 percent to 300 percent vegetative growth, depending on the type of plants and the quality of the soils; poor soils will likely produce higher performance gains than good soils. Positive results can also be seen for fish and livestock fed the leaf protein and biochar nutraceuticals. Poultry can freely range the alleys to the benefit of both plants and animals. Grazers can be moved through rotational pastures that take advantage of water reservoirs and high-quality supergrasses. Gourmet mushrooms such as morel (*Morchella* var.) can be pre-blended with the biofertilizer and seasonally harvested.

Growing nutrient-dense, no-till, organic food and perennial fibers on marginal lands, using bioenergy and biofertilizers, creates a resilient, circular bioeconomy. Like Mother Nature, these systems waste nothing. Nothing need leave the system. Neither raw material nor pollution — representing the depleting wealth of the land — leave the system. What does leave are high-value, locally produced products — providing return on social capital invested. This is the "Cool Farm."

Transportation presents an energetic challenge in the new solar-based carbon world. Nearly all modern forms of transportation evolved in an era of cheap fossil energy and diminish in economic viability when costed on renewable sources and life cycles. Diesel-powered semi-tractor-trailers and locomotives will need to be reimagined and likely replaced or retrofitted. There could be new generations of electrified towpaths for barges and gondolas, magnetic levitation rail, and other innovations, but these costly innovations may be fragile in an era marked by overpopulation, resource constraints, and economic contraction. They are unlikely to provide a stable foundation for local commerce. We could see the return of barge, sail, and animal-powered transport.

If taken to global scale — rotational planting an area the size of India each year, converting to Cool Farms and installing Cool Labs in every village — the price of captured carbon per ton would drop from more than $200 to $20. Moreover, while most other forms of carbon capture and storage require uncompensated operating and maintenance costs, the Cool Lab is immediately profitable. It represents continuous and adaptive economic development — antifragile profits, not ongoing expenses — as well as gains in ecological health and biodiversity. It ticks the boxes for sustainable development.

In contrast with the forty-five-year gradual expansion of soybean cropping from the early 1960s to reach 200 million hectares (772,000 square miles) today, Cool Farms, employing integrated agroforestry, offer up to five times the protein per hectare of soy while providing a far greater, and more immediate, return on financial investment.

Cool Labs leverage the existing financial and technological landscape of the world today but reimagine the way products are produced. The number of possible cascades is limited only by the imagination, skills, and willingness of local residents. We are at the dawn of a new kind of lean, clean, nature-centered economy. The entire system heals the earth, rebalances carbon, and generates economic security and resilience for more people.

Carbon cascades can turn almost any human settlement into an ecovillage, although the criteria for what defines *ecovillage* must necessarily include a few more elements than merely having a Cool Lab or permaculturally designed support systems. Ecovillages are based on a cohesive worldview, an abiding respect for the ecological integrity of your home, a circular local economy, and a culture of peace and mutual respect. Depending on your starting point for each of these elements, bringing all of them into harmony can take time and effort.

We asked former NASA engineer Frank Michael to give us the drawdown numbers from the energy and food production system we've outlined, taken to global scale (covering 20 global hectares of low-productivity land[5]). Michael said we could sequester carbon from the atmosphere at the average rate of 8 GtC/yr over the first eight years, and reach 13.6 GtC/yr by twenty-four years.[6] We would achieve the cumulative storage of the 600–700 GtC required to bring atmospheric carbon back to preindustrial levels within about fifty years, taking into account the oceans' CO_2 outgassing feedback and other black swans. Carbon would be stored in the world's soils and living biomass and could therefore provide additional benefits beyond sequestration. Were nations to collectively phase out fossil fuel emissions, the reduction targets would be achieved much sooner. It is possible to recover the Holocene.

In April 2018, Hans-Peter Schmidt, whom we met previously, joined four other authors on a paper published by *Environmental Research Letters* titled, "Biogeochemical Potential of Biomass Pyrolysis Systems for Limiting Global Warming to 1.5°C."[7] The authors showed that by adding two other cascades to the biomass-to-biochar-plus-energy process — bio-oil (pumped into geological storages) and permanent-pyrogas (capture and storage of CO_2 from gas combustion) — the land requirement dropped by up to 60 percent, while the benefits from yield increases could, in time, diminish land requirement by another 3–38 percent. The equivalent drawdown volume of greenhouse gases Michael had calculated would require 10 billion hectares (38.6 million square miles). Equally careful as Michael to observe biodiversity and cultural guardrails, Schmidt and colleagues estimated to only need 82–362 million hectares. Three hundred sixty-two million hectares (3.6 million km^2, or 1.4 million square miles) may seem like a lot of land, but we know from standard reference studies that there are at least five times that available in low-productivity, marginal lands in need of soil revitalization and ecosystem regeneration.[8]

Adding one percentage point of soil organic matter means that around 27 metric tons of organic matter per hectare (12 short tons per acre) enter the soil and remain there. Around two-thirds of organic matter added to agricultural soils will be decomposed by soil organisms and plants and given back to the atmosphere as CO_2, CH_4, and N_2O from that metabolic process. To add permanently 27 tons, a total of 81 tons of organic matter per hectare (33 per acre) is required. This cannot be done quickly or it just washes or evaporates away and can overwhelm ecosystems. A slow process is required.

An example of how this could play out in Costa Rica, Haiti, Africa, or anywhere else can be seen in the Loess Plateau of northern China where fertile soils were overworked until they had to be abandoned. At the time of abandonment organic carbon concentrations had dropped to under 3 percent. Thirty years later Loess soils had regained concentrations of 6 percent by persistent human effort to aid natural processes. If natural restoration were accelerated by amending soil carbon in both metabolizable forms (including crop litter, manures, food waste) and recalcitrant forms (such as biochar), the process picks up speed. This could happen virtually anywhere.

Coral reef scientist Thomas Goreau told us that at typical current levels of carbon farming it would take thousands of years to draw down excess CO_2, but biochar and enhanced mineralization could draw it down in as little as decades. If the recuperation of soil carbon became a central goal of agricultural policies worldwide, it would be possible and reasonable to set as an initial goal the sequestration of ½ ton per acre-year (1.5 t/ha-y or 500 grams per m^2-y).

A farm that switches to organic, animal-powered no-tillage methods can sequester 1–4 tons of organic matter per acre (~2.5–10 per hectare) per year. By employing perennial polycultures, rotated pastures of grazing animals, trees, and wild plant strips, that amount can be doubled or tripled. As soil conditions improve, erosion and pests decline, and the land comes back into balance, our target goal could be increased, up to a point at which the land itself responds, telling us that further gains are unlikely or marginal. Farming this way globally could sequester about 8 percent of the current total annual human-made emissions of 10 gigatons of carbon. However, the fertility gains (equivalent to more than all current global fertilizer production) would mean that chemical fertilizers could be (and should be) eliminated where this style of carbon farming is practiced. By reducing emissions of nitrous oxide from fertilizer (equivalent to approximately 8 percent annual human-made green-house gases) and the transportation and energy impacts of fertilizer production, we shave some additional percentage points off global emissions.

Even a modest start, such as by elevating the soil carbon content of existing farmed soils by 0.4 percent, could have the potential to offset global greenhouse gas emissions by approximately 20 percent per year. By mid-twenty-first century, we could increase the total world reservoir of carbon in the soil by two percentage points, taking that much away from the atmosphere and oceans. In this way it is conceivable to restore our soil carbon reservoir to 10 percent and even to regreen and reforest equatorial deserts and, in so doing, end the climate crisis.

More importantly, it can pay for itself and go anywhere and the social cascades can have as big a payoff as the economic and environmental bene-fits. From very modest beginnings, like Art Donnelly going to Costa Rica to help the children of coffee farmers in 2009, carbon cascade projects can quickly spread. In the next chapter we describe the work of Govardhan Ecovillage in India, the Maya Mountain Research Farm in Belize, Xl'a Ká Vergel in Mexico, and El Valle in the Dominican Republic. All of these are in the formative stages of community-led, microenterprise-driven carbon cascades to reverse climate change.

CHAPTER 19

Grassroots Solutions

If you have built castles in the air, your work need not be lost;
that is where they should be. Now put the foundations under them.

— HENRY DAVID THOREAU, 1854

C onversations about "development" are often phrased in words like *growth, jobs, stock market highs* and *lows, gross domestic product,* or *trends in consumerism.* Some of the more farsighted use metrics such as inclusion, intergenerational equity, longevity, marriage stability, and happiness. Yet, just as all politics is local, all economics are personal. It comes down to how well any community — be it a rural cluster of farms or an urban neighborhood — fends for itself in the volatile world of the twenty-first century.

Solutions that endure usually begin at the bottom, building regenerative, circular economies based upon local assets — human and natural. They care for all, protect the planet, and reach out to help their less fortunate neighbors. They are organic, resilient, and antifragile. Here are a few of their stories.

Gauranga Das at Govardhan (India)

Today half of the world's population lives in urban areas. By 2050 the proportion in India is expected to be 80 percent. Three-quarters of India's 83.3 million rural villagers earn less than five thousand rupees ($78) per month. Half do not own land. Those that do are often indebted to banks for equipment, fertilizer, and pesticides, charged interest rates they cannot pay. Suicide rates in the countryside are double those in urban areas. A few years ago, two hours north of Mumbai, Shri Gauranga Das established

Govardhan Ecovillage and its philosophy of "symbiotic recycling" — a merger of science and Vedic teachings that integrates organic farming, biogas, and green buildings into a circular local economy.

Organic fertilizers, compost, and mulch are produced locally at practically no cost. Biogas replaces wood or gas for cooking. Construction wastes like broken cement are used in raised bed gardens, in cob houses, and as infill for infrastructure. Green buildings of compressed stabilized mud bricks are cool in hot weather and warm in cool weather. Rainwater management provides irrigation in dry months and recharges aquifers from monsoons. Graywater and blackwater flow to bioreactors that use plants, earthworms, and aerobic microbes to remove suspended solids, pathogens, and odor, returning energy and fertilizer.

India has long been one of the leaders in biochar, thanks in no small part to the work of Dr. N. Sai Bhaskar Reddy Nakka at the Appropriate Rural Technology Institute in Phaltan, a short distance south of Govardhan. For more than twenty years, Reddy has been taking biochar compost blends to farmers, making biochar bricks for green buildings, using biochar powders for waterless cleaning, and designing efficient home stoves. Worldwide, the three-stone open home fire is currently responsible for more childhood deaths than malaria — eight million last year. Reddy has personally trialed more than fifty designs of low-cost gasifier cookstoves for homes and businesses.

Now Govardhan Ecovillage is passing its symbiotic practices to sixteen nearby tribal villages. Four hundred families have come together to plant more than one hundred thousand food, forest, and medicinal trees that will absorb 2,000 tons of CO_2 as they grow. Das calculates that if even 1 percent of India's villages follow the model of Govardhan, 4.7 million tons of CO_2 will be drawn down annually.

Christopher and Celini Nesbitt at Maya Mountain (Belize)

From Punta Gorda, the capital of the Toledo District of Belize, you ascend the foothills of the Maya Mountains toward the pyramid ruin of Lubaantun near San Pedro, Colombia. This was a late Classic Maya ceremonial and commerce center where the famous crystal skull was found by the teenage daughter of archaeologist F. A. Mitchell-Hedges in 1926. From a thatch-roofed village the journey is by river, as a boy with a dugout "dory" cedar canoe takes you an hour upstream to a shallow bend with tall stands of bamboo.

The river's source is a massive spring that bursts from the ground a quarter mile farther — from a vast underground river system that drains the 100,000-acre Columbia River Forest Reserve, a uniquely pristine natural area of broadleaf tropical forest, sanctuary to howler monkeys, jaguars, monarch butterflies, and birds of paradise. The landscape is strongly karsified — riddled with deep caves and some of the largest cenotes in the country (one is a quarter mile [0.4 km] wide). Shallow caverns of quartz-rich rocks provide breeding habitats for diverse and secretive animal populations.

Christopher Nesbitt decided to buy his piece of this paradise in 1988. After leaving Antioch College at nineteen, he took a job as an agroforestry extension agent at Toledo Cacao Growers Association. Everything he learned about cacao, agroforestry, self-sufficiency, and self-finance he put into the Maya Mountain Research Farm. As we walked the land, he paused in the shade of a large avocado he planted in 1989.

"More avocados than can be eaten by one family," he said. The same is true of his mangoes. He started a piggery to feed the pigs his surplus fruit. He wants to use their manure to make methane for his kitchen. He has built a tank and pond aquaculture system, although most of the fish in his kitchen still come from the river or the Caribbean Sea.

The heart of Nesbitt's method is Inga alley-cropping, a system of deep mulching using pruned green leaves from nitrogen-fixing Inga trees contour-planted in hedgerows. The system began in the mid-1980s with trials by the University of Cambridge, Kew Gardens, and others followed by a pilot program with families in Honduras. The trials and pilot proved that fuel and food security can be quickly achieved for forest families with the added benefit of virtually no need for weed control.

After taking a permaculture design course in 1991, Nesbitt dug swales across his hillsides and added a number of ground-hugging plants and vines to keep the soils shaded and protected from erosion. For him, cacao was the keystone plant in the system, and there was good reason that the Maya placed a high social value on it, beyond its health and nutritional qualities. Cacao's scientific name *Theobroma* means "food of the gods."

Raw cacao beans contain magnesium, copper, iron, phosphorus, calcium, anandamide, phenylethylamine, arginine, polyphenols, epicatechins, potassium, procyanidins, flavanols, and vitamins A, B, C, D, and E. Long before the Swiss and Belgians became famous for their chocolate confections, the ancients mixed cacao with maize, chili, vanilla, peanut butter, and honey to make various beverages and confections. The Aztec and Maya cultures used the beans as currency, a practice that persisted in the Yucatan until the 1840s. Given world prices of US $2,250 (industrial grade) to $8,000 (fair trade organic) per metric ton, the beans are a form of currency still. When Mayan women go into labor, they are given a big thick mug of toasted cacao, cane sugar, and hot water. Because it is rich in calories and healthful, that big mug can see them through days of labor and the recovery afterward.

On the stones outside the kitchen, under the roof, and out of the rain, Christopher and his wife Celini have a bowl of cacao beans fermenting. They are left there for a week and grow a fine white spiderweb of hyphae as they incubate. They didn't need any starter; the airborne yeasts did the job. After seven days, they are rinsed, ground, and toasted. This year one of Chris's permaculture students, Claudia Gonzales, brought a cacao grinder for the farm, assembled by a new branch of Bicimachina in Mexico. It is a modified recumbent bicycle that lets you grind many kilos of cacao in a short amount of time. Chris is reveling in the wow factor this will have when his neighbors see it.

You could live quite comfortably on the breadnuts, avocados, corn, bananas, coffee, fish, beans, and all the rest. You could drink from the river, although the Nesbitts harvest water for the kitchen from a spring farther uphill. If you glance around their open-air kitchen, you'll see purchased cans and jars containing items like powdered milk, granulated sugar, olive oil, foreign teas, iodized salt, and baking soda. These are all part of a Western diet but not indispensable here. Successive civilizations did just fine without them.

Western civilization was founded on savannah grasses, irrigation, and the plow, and, like our religious traditions, the agriculture we are accustomed to is a five-thousand-year-old relic that grew surpluses of grains, but also bequeathed enormous and spreading deserts, centralized and hierarchical wealth systems, standing militaries, and a seemingly intractable global ecological crisis.

No green chlorophyllic cells can photosynthesize 100 percent of the sunlight that falls on an unfiltered square inch of ground in a day, so most of that solar energy is bounced back to space or lost to heat. Multistoried polyculture forests with climbing vines and ground covers, on the other hand, share dappled rations of light as a community and have far greater absorption, production of oxygen, retention of nutrients, and potential to provide food.

The Nesbitts' near-term pioneer crops are annuals such as corn and beans, or pineapple, pigeon pea, squash, and melons planted between the corn contours, along with perennials such as nopal cactus, yam, purslane, basil, amaranth, and gourds. The intermediate crops are perennials such as avocado, golden plum, zapote, sea almond, allspice, bamboo, palms, breadfruit, coconut, coffee, coco-yam, banana, citrus, mango, cacao, papaya, tea tree, euphorbia, noni, blackberries, gooseberry, chaya, ginger, and pineapple. They will yield sweet fruits, jams, wines, basket fiber, soaps, beverages, and medicines after a few years of fast growth. The long-term crops are samwood, mahogany, cedar, teak, Malabar chestnut, sea chestnut, and other slow-growing trees that will close the overstory.

An important feature of the tropical landscape design is the creation of soil. Here in the equatorial latitudes, much of the nutrient value of soils is carried in the standing plants, and the process of transmitting soil elements through decomposers and carriers to next year's crops is very fast. The couple make a daily ritual of making biochar in a special gasifying stove they use to cook breadfruit. The biochar is mixed with animal manures and spread below the food trees, which then grow faster and fruit sooner.

Lately the Nesbitts have been thinking about how the small town just downriver is starting to outgrow its borders and infringe on the mixed-used forest and its traditional caretakers. Chris sees the possibility for strategic intervention. In 2017 the Nesbitts began working on a proposal for the Common Earth project being developed by the Commonwealth of Nations, of which Belize is a member. Common Earth wants to channel finance to worthy projects that focus on regenerative development to reverse climate change. The Nesbitts' plan is to contract with local forest smallholders to supply feedstocks to a Cool Lab, centrally located within the village, from the waste products of their milpas. They could also supply raw superfoods such as cacao, coconut, vanilla, moringa, acai, goji berry, and blue-green algae for processing into high-value products.

The Nesbitts have a pretty good idea of where they might site a new village school and health clinic. With biomass-to-biochar energy and quick returns on investment they envision opportunities to offer gainful, anti-fragile employment to the youth of the village as they finish school and look for work. Given just a little assistance to get started, they will show Belize, the Commonwealth, and the world how it can be done.

Bernardo del Monte in Xl'a Ká Vergel (Mexico)

A few hundred miles north of the Maya Mountains is another experiment in integrated agroforestry begun in Southern Mexico by Bernardo del Monte, née Bernd Neugebauer. Albert first met Bernardo some years ago when asked to work on the permaculture design for a planned ecovillage on Mexico's Caribbean coast, Puertas — the Portal.

Albert visited Bernardo in his modest thatch dwelling where he is rein-venting civilization. At Xl'a Ká Vergel, he is reestablishing the indigenous pattern of that region in all its dimensions — hydrological, physical, social, and spiritual. Bernardo said, "Humanity has been around a long time and we have had tremendous successes, but certainly not this present civiliza-tion. Others have been much more successful in many, many ways. None of them have ever been complete. There has always been another task and another task. We've just gone through one very specific task-solving period. We'd better look at the experience of the whole civilizing process and pull the threads together.

"If you look at two civilizations, the present civilization and the Maya civilization, and contrast the two, that helps you define the steps forward. It is not about recovering the Mayan, it is about being able to see the array of options. So my question is really, what do we have to do to make civiliza-tion work?"

Del Monte has been working on a manifesto, called variously the New Deal or the Second Civilization, and his petri plates are Xl'a Ká Vergel and Puertas. His idea is to offer affordable homes to all socioeconomic groups, with the caveat that anyone choosing to join will have to abide by the sus-tainability guidelines — restrictive covenants — that may include, among other things, ecological codes for buildings and businesses, a circular, waste-free economy, and requirements of public service. The village will have only one social class, recognizing that interests, needs, capabilities, and routines of inhabitants do not always converge but may vary.

Similar efforts have been going on around the world since architect George Ramsey first coined the word *ecovillage* in the early 1970s. Today there are some twenty thousand examples in more than sixty countries, and the idea is gaining momentum. China plans to sponsor one hundred new ecovillages in the next five years. Senegal envisions more than ten thousand.

While the exterior views of these villages may be quickly recognized by their organic gardens and landscapes, natural buildings, wind farms, and solar panels, what lies just beneath the surface is an invisible architecture far more profound. The heart of the next civilization is borne in the hearts of these explorers. As we walk the paths where his first new village is emerging from the forest, Bernardo explains what we are seeing. "Lessons from vernacular Mayan architecture give us clear indications of how this process has worked in the past," he says.

"The city starts from these. Designers edit the first steps — proposals for thatch-roofed pods of a few homes and their shared infrastructure. These pods are to become autonomous units with their own energy supplies and eventually their own gardens. Their gray- and blackwaters, compostables from the kitchens, and the biochar from their stoves or kilns will be recycled to build the fertility of the soil."

He first learned about biochar from the University of Bayreuth soils scientist Bruno Glaser, an early discoverer of pyrogenic carbon in the terra preta soils of the Amazon. Del Monte has been making and applying biochar ever since, changing the soils of Xl'a Ká Vergel from a chalky ocher paste into a dark granular sugar. The photosynthetic capacity of his land is astonishing. When the rains come, the land absorbs the water and thrives. Trees that look decades old are actually only three or four years old. We walk up a hill to where he is building a pod overlooking a water system that he reconstructed from the traces of the way it had been done five hundred years earlier. Based on his experience with the Maya communities of the Yucatan, Bernardo has narrowed his formula to four steps:

1. Start from a true perception of reality.
2. Prioritize what needs to be changed.

3. Define the simplest possible first step to transformation and implement it.
4. Motivate.

"You need to implement the second step and repeat the sequence five times," he told us. After five successes motivation grows into enthusiasm, which is the strongest engine of change.

"Enthusiasm is what you experience when falling in love. You become capable of doing anything possible or even impossible to obtain your object of desire. When we deal with the Earth and the feeling of enthusiasm it arouses, it is as if we fall in love with the Earth — which is exactly what both people and the Earth need!"

Santiago Obarrio and Carolina Chicero in El Valle (Dominican Republic)

As Santiago Obarrio tells it, when he was twenty-eight, he decided to escape to a mountain, to live how he thought humans should. He grew his own food and lived a happy existence, but felt unfulfilled. He told us he couldn't bear the thought of saving himself and not helping society. One thing led to another and eventually he found himself being offered a master-planning opportunity for a remote valley on the north coast of the Samana Peninsula.

The area was remote for good reason. Steep mountains ringed the valley, and in the rainy season the river rose and cut off the only entrance pass. At those times the valley could be reached by sea, but if the waves were high, it might mean approaching in heavy surf with no docks or easy landing areas. There were a number of farming and fishing families living in the valley, but the pattern of exodus was destructive. When a child reached high school age, they had to relocate to a city outside the valley. Few returned after they finished school. While the valley was rich in natural beauty, it was poor in human resources and amenities. Yet all this was about to change. The government planned a new highway and a bridge over the river. Development would be coming to El Valle. To Obarrio, this seemed both a hazard and an opportunity.

Obarrio enlisted a few friends to join with him in a business partnership for the purpose of developing an ecological plan for the valley. They called the business Qi. Their business plan consisted of three elements — Cool Farms, Cool Lab, and Cool Village. His partner, Carolina Chicero, told us, "Watching how other developers are slowly destroying paradises with 'eco-development,' especially in the Dominican Republic, we decided to take this example as far as we could envision. More than thirty experts came to design this ecoregion. The 3,000 hectares [7,400 acres] are a watershed sanctuary where six defined ecotones interact, creating astounding biodiversity of trees, birds, turtles, and soils." It is little wonder the local residents can live on fishing and agriculture with little need for the outside world much of the year.

"We are currently working with the local community and building our own lodge, which will be sustained entirely by products coming from El Valle. We use the mineral water from the land, leave no wastes, and plant more food. We have already regenerated a spring and are harvesting fresh-water fish and shrimp."

The company brought in an ecovillage architect, Greg Ramsey, to design the master plan for the 3,000-hectare district. Ramsey convened valley residents, spoke with the farmers and fishing families, and together they worked up a code of covenants and restrictions. Cars will be kept at the periphery. Within the district, marked by an entrance gate and transition point, transportation will be restricted to slow microvehicles, including motorized bicycles and tricycles, hoverboards, and electric carts that can operate within the pedestrian paths. Gradients are restricted and the speed limit is 15 miles per hour (24 km/h). Landscaping must comply with standards for organic land care and the native plant species list. Development will be held to a 90:10 offset density ratio, meaning that only 10 percent of the land area of the valley will be permitted to be developed. The remaining 90 percent will be kept wilderness or mixed-age, mixed-species ecoforestry.

Lacking deep-pocket investors, the company pulled itself up by its own bootstraps with its Cool Farms. Assisting local farmers to do what they knew best, Qi introduced organic, regenerative methods such as biochar-compost blends, compost tea inoculants, mineralization, and hybrid seed for fast-growing plant strains. Leading with this allowed Obarrio to get going at minimal expense but maximum impact. From the carbon fixation point of view, the drawdown is better than 10:1 for solar or wind power, and the energy return on energy invested is better, too. The internal rate of return is 90 percent, ten times the usual rate of return for solar power.

"We expect to cover our operating costs for the company in the Dominican Republic for the next eight to ten years with this operation alone," Obarrio said. "Although there are many products we can produce after we fix carbon from the air, we have decided to focus on the most essential commodities and the ones that can virally grow inside huge markets. One hundred and fifty dollars invested in a Cool Farm erases the carbon footprint of an individual for ten years and can double your investment in four to six years."

The first of QI's commodities are biofuel feedstocks made from hybrid perennial grasses produced at net drawdown — Cool Fuels. For the clients who are purchasing these, Cool Farms are able to meet their needs for 33 percent less than what they had previously been spending on fossil fuel. Second-tier commodities, planned for 2020, will be nutrient-dense Cool Foods, both those that grow in tropical valleys, and proteins and medicinals extracted from leaves of biomass crops or seaweed before they are used to make electricity and biochar.

According to the World Economic Forum, if public-sector investment (governments and international agencies) was increased to $130 billion and more effectively targeted, it would eliminate the investment gap between the Paris climate goals and actual drawdown projects by mobilizing around $570 billion in private capital.[1] Cool Farms illustrate how that could happen.

After their first season, which included being battered by Hurricane Irma, Obarrio's Cool Farms, now 100 percent organic and massively making and applying biochar, were able to show a 33 percent increase in soil carbon. The company's projection is an average drawdown rate of 10–20 tons of CO_2 per hectare per year.

Once their lodge is complete, the couple plans to construct some fifty cabins for guests and vacation home owners. Wild side ecotourism will help pay the way for cool development of the ecodistrict. Obarrio says, "Since our energy source is carbon negative (microhydro plus gasifier and generator), if a guest turns on the LED lights he will be cooling the planet. His trip will become carbon negative by adding biochar to soil with waste wood from the forests turning it into fertilizer for local producers. This is how we are teaching locals not to cut their trees but to value the forest as their source of high yields for their farms."

The other element that can help pay the way is phase three — a village Cool Lab. Qi is taking its profits from the Cool Farms and other projects and, rather than selling raw farm commodities to outside buyers, will develop value-added cascade products. Obarrio says he expects to pay the early investors around 24 percent profits per year on the first model. After that, the company says it can use El Valle as a showroom to help others construct Cool Labs at scales of anywhere from $1 million to $100 million. It may even help finance them by offering Cool Bonds.

Obarrio told us, "We are open source. We need this to go viral. The fight between economic growth and sustainability starts from a mistaken premise. Regenerative systems can fuel economic growth. The only difference in our development is that this time carbon must come from the atmosphere and go under the soil, cooling the planet rather than warming it. It is very simple, but also very revolutionary. It will change the lives of the poor, the rural people, those who have suffered from the old way. It will energize the protection of nature."

––––––––

These are four brief case studies — from India, Belize, Mexico, and the Dominican Republic — and we could easily have included many more. Each of these personal stories begins with something akin to what Joseph Campbell called the hero's journey, a recurrent theme in myth and legend. The protagonists — Gauranga, Christopher and Celini, Bernardo, Santiago and Carolina — set off to part company with their pasts. They underwent rites of purification. They emerged transformed and inspired to undertake a Herculean task. They dedicated their lives to the quest.

In each of these stories our heroes started young and did not squander their youth. The tools they picked up and learned became the modern weapons of carbon drawdown: forests, mangroves, renewable energy from waste, biochar, water gardens, ecovillages, and Cool Labs. They each progressed from thinking of their own betterment to planning, teaching, and constructing at the scale of the village, bioregion, and nation. The responsibilities they decided to shoulder steadily grew throughout their lives, as if they at some point had said, "More. Please. Just lay on a little more. I am that strong," and then they grew that strong. In the course of writing this book we came across many other stories — from Borneo, China, Africa, Latin America, and Eastern Europe — that inspire us and instill a great sense of hope. They redeem our faith in human ingenuity and compassion.

CHAPTER 20

Civilization 2.0

The outer boundary of what we currently believe is feasible is still far short of what we actually must do. Moreover, between here and there, across the unknown, falls the shadow.

—AL GORE, 2007

In *The Wizard and the Prophet*, Charles C. Mann wrote, ". . . [I]t is terrible to suppose that we could get so many other things right and get this one [climate change] wrong. To have the imagination to see our potential end, but not have the cultural resources to avoid it. To send humankind to the moon but fail to pay attention to Earth. To have the potential but be unable to use it — to be, in the end, no different from the protozoa in the petri dish."[1]

The only way to endow the future with a chance of reversing climate change is to transition as rapidly as possible to a habitation pattern (an economy) that does not push carbon into the atmosphere and oceans but draws it in. That change, the late economist David Fleming urged, "will depend for its existence on a deep foundation in culture."[2] It cannot concentrate to scale. It must spread.

Fleming predicted that it won't be hard to move away from our market-based, resource-to-pollution society; it will fall away so fast that we will find it hard to believe it was ever there. "The task, on the contrary, is to recognize that the seeds of a community ethic — and indeed, of benevolence — still exist. It is to join up the remnants of local culture that survive, and give it the chance to get its confidence back."

As we confront the hyperwicked problem of human-induced climate change, recognizing as we must that "human civilization is a heat engine,"[3] and then we try to layer onto that challenge our urgent needs to bend the curve on human population, achieve the Sustainable Development

Goals, and design a transparent, nonviolent, fair, and equitable economic regime going forward that honors the rights of nature — all this can easily overwhelm. In the architectural design process this is called "the whole systems challenge."

One indispensable technique is patience. We need to remember that understanding only arrives from the frequently messy process of bringing preconceptions to the fore, examining them, discarding some, and accepting others. Emotions play a role. Skilled facilitation helps. Complexity is inescapable. To reach a satisfactory or, at least, acceptable settlement, one needs to be prepared to "sit in the fire" — to appreciate the interconnectivity of discordance and decision. We cannot exclude elements that bear on the puzzle simply because they are messy. We must learn to value fuzziness for what it can bring to a cohesive design that can endure.

The Buddhist philosopher Thich Nhat Hanh compared this acceptance of complexity to a sheet of paper:[4]

> If you are a poet, you will see clearly that there is a cloud floating in this sheet of paper. Without a cloud, there will be no rain; without rain, the trees cannot grow; and without trees, we cannot make paper. The cloud is essential for the paper to exist. If the cloud is not here the sheet of paper cannot be here either. So we can say that the cloud and the paper inter-are. "Interbeing" is a word that is not in the dictionary yet, but if we combine the prefix "inter" with the verb "to be," we have a new verb, inter-be. Without a cloud we cannot have paper, so we can say that the cloud and the paper inter-are.

Daniel Wahl, in *Designing Regenerative Cultures*, writes of this passage:[5] "In many ways, the word 'interbeing' describes a shift in the perception of self and other that lies at the heart of co-creating regenerative human cultures and a sustainable human presence on Earth. Transformative innovation for regenerative cultures drives the shift from an industrial growth society, based on extraction and exploitation of natural resources and informed by the 'narrative of separation' to a life-sustaining society, based on regenerative agricultural and industrial processes and informed by the 'narrative of interbeing.'"

Seen as a complex, adaptive, and dynamic system, any human community shares common features not only with other communities, but with flocks of geese, the Earth's magnetic field, or a river. Wahl says they each:

- are nested open systems
- have fuzzy boundaries
- change their structure and behavior over time
- contain feedback loops that create nonlinear effects
- are shaped by unobservable, but influential, emergent structures, and properties

These properties are all advantages, even though they can create obstacles and frustrations to organizers. To solve problems, pieces cannot be isolated. They have to be approached as wholes. To do this, many of the innovators we've mentioned throughout this book have developed interactive methods — games, stories, group exercises, and more — that extend boundaries enough to encompass these wholes.

The property of interbeing is the driving characteristic of the cascades we have been describing. What makes the daily income of low-tech, easily sourced, sustainably produced biomass — photosynthesized sunlight — able to supplant a onetime, fragile, million-year savings account of fossil energy, even without government subsidies or carbon credits, are carbon cascades. The same feedstock provides multiple products and services, and those same products and services provide multiple cycles of use and reuse, each adding more value to the process as a whole. Beyond the value to the economy, or the business case to be made, there are compounding social and ecological cascades of equal, if not greater, importance to accomplishing the larger mission, which is to sustain the still fragile human experiment.

In his 2006 State of the Union Address, the president of the United States, a Texas oil man, chastised his countrymen, saying, "America is addicted to oil." Then George W. Bush offered his solution: "By applying the talent and technology of America, this country can dramatically improve our environment, move beyond a petroleum-based economy."

There are cures for addictions, but the struggle for any addict is to produce permanent change. Yes, humans will always advance their technologies, but if we expect to get out of the climate dilemma without going extinct there is only one biophysically verifiable path. We have to engineer an industrial decarbonization glide path that follows an 11–20 percent decline slope in greenhouse emissions for the next eighty or more years, not just to zero, but beyond. We have to raft through the cascades into negative emissions territory – the land called Drawdown – by 2030 if we can, but mid-century at the latest. Getting there somewhat later is not an option. The party will be over. We would be the final progeny of a one-hundred-thousand-generation run for one magnificent mammal.

Ideas have ways of graduating from a time spent in the experimental laboratory to early, tentative, real-world trials, to full-blown adoption (especially in times of crisis and change) as new social norms. When they succeed, they linger. If they fail, they are replaced. We are in such a period of crisis and change now. Business and opinion leaders are making the rounds of the laboratories in search of ideas and business incubators to identify sufficiently mature concepts to move to center stage. Steady-state economics, B-corp public banks and credit unions, permaculture, regenerative design, natural climate solutions, microenterprise hubs, and ecosystem restoration are all in the anteroom, perfecting their pitches. Many of these have been profiled through the course of this book. Each may have a hand in writing humanity's future.

Admiral Hyman G. Rickover, tasked by his country to nuclearize the navy after World War II, chafed under the constant harping by government oversight committees, the Congressional Budget Office, and others who thought they would find waste and corruption in his program. He told them:

> An academic reactor or reactor plant almost always has the following basic characteristics: (1) It is simple. (2) It is small. (3) It is cheap. (4) It is light. (5) It can be built very quickly. (6) It is very flexible in purpose. (7) Very little development will be required. It will use off-the-shelf components. (8) The reactor is in the study phase. It is not being built now.
>
> On the other hand a practical reactor can be distinguished by the following characteristics: (1) It is being built now. (2) It is behind schedule. (3) It requires an immense amount of development on apparently trivial items. (4) It is

very expensive. (5) It takes a long time to build because of its engineering development problems. (6) It is large. (7) It is heavy. (8) It is complicated.[6]

The future we outline with this book will likely resemble Rickover's program. It is behind schedule and it is complicated. To rebalance the carbon cycle, we need to increase the sinks and decrease the sources. Looking at the sawtooth pattern of the Keeling Curve, we want to shorten the back edge and lengthen the lead edge of each tooth. Let's not be naive. Displacing dirty old fossil carbon with clean new carbon in soils, supercapacitors, and stormwater management systems cannot, on its own, reverse or rebalance humanity's cumulative or even current carbon misallocation. There is no silver bullet when it comes to climate change, and no single strategy will suffice. Our carbon buckshot approach, maximizing drawdown through cascading processes and products, can go much further than might have been thought, but still cannot get us all the way.

If you take the emissions from fossil fuels, cement production, land use change, and other human influences, count how much of that has been added to the atmosphere and how much to the oceans, and then account for the natural removal processes, if emissions stopped today, we'd need to pull about 650 gigatons of carbon to put the atmosphere back in proper balance for human existence.[7] Applying cascades to the various industries and products described in this book could remove, by our back-of-the-envelope calculation, 50.6 gigatons of CO_2 or equivalents among other greenhouse gases per year. That number exceeds present human-made

Table 20.1. Carbon Math: Selected Additional Sources

Source	GtC
Seaweed	2.6
Crop residues	0.5
Sludge	0.1
Other*	0.3
Total	**3.5**

Note: Pyrolysis sources in gigatons of C per year.

Table 20.2. Carbon Math: Total Additional Sinks

Industrial product	Drawdown potential (GtCO$_2$-e/y)
Construction aggregates*	31.9
Sand	15.0
Asphalt	1.8
Soils	1.7
Other†	0.2
Total	50.6

Note: Potentials for atmospheric removal in gigatons of CO_2 per year

* Crushed stone, dimension stone, bricks, and rubble

† Diatomaceous and fuller's earth, clay, kitty litter, wastewater filters, carbon black in tires, and activated carbon

emissions (~39 gigatons/yr) by 10 gigatons, meaning that in combination and at full scale they could pull that much CO_2 from the atmosphere and oceans annually. We could restore a safe climate even faster if emissions were to decline as steeply as the Paris Agreement requires.

Of course, emissions are not stopping. They haven't even been slowing down on a global level. Every year that situation persists, you can tack on more that will have to be removed later. Much more, as these small additions cross thresholds that trigger tipping points, positive reinforcing reactions like melting permafrost and droughts in the Amazon.

To rescue the human prospect from the climate event now brewing, we will need to radically change everything about how humans live on Earth. We have to go from spending carbon to banking it. We have to put back the trees, wetlands, and corals. We have to regrow the soil and turn back the desert. We have to save whales, wombats, and wolves. We have to reverse the way carbon is going, change direction and send it in exactly the opposite direction: down, not up. We have to flip the carbon cycle and run it backward. We'll need civilization 2.0.

Halfway measures won't work. There are no magic wands, no fairy dust. There is a hard reality. Either we face it, or we die, and with us goes everything we did — all the poetry, art, and music; the great books and plays; the marble buildings, museums, libraries, and moonshots. They will be as mists of the morning, and then they will be gone — a thin carbonifer-ous layer in the earth, marking the Holocene-Anthropocene boundary.

We offer here a prescription: Planting enough forests while simultane-ously curbing emissions will get us to net drawdown territory; those parts per million of carbon in the atmosphere will begin to go from 410 to 405, then 400, then 395. Make biochar an integral part of that forest-indus-try-energy mix, and start to sequester it in concrete highways and high-rises. Grow kelp and, after pressing it for leaf protein, char that and build coral-restoring coastal filter barriers. Carbon abuse and waste become carbon abundance and recycling. Now the change we make starts to stay changed. Circular carbon economies begin to cascade.

Many proposed climate change mitigation strategies — BECCS, DAC, geoengineering — come at enormous costs and carry great risks. Carbon cas-cades, however, provide companies and countries with cost savings. Many are cost-neutral from the start when reduced tipping fees are factored in. Others can become profitable by harvesting the heat of carbonization. Cascades go far beyond "just" carbon rebalancing. They can regenerate landscapes, communi-ties, and ecosystems. Profit making and planet saving is not a zero-sum game.

This book is meant to be a love story. Right now carbon is getting a bad rap. Carbon creates dirty energy. Carbon creates grit, grime, and gunk. Carbon should be global warming enemy number one. But in truth, carbon is something we should all love and cherish. Carbon is life. In the right bal-ance, carbon gives life.

The five most common elements in the universe are hydrogen, helium, oxygen, carbon, and neon. Because helium and neon are typically inert, it remains for the other three to be the building blocks of life. Carbon and hydrogen make up only a fraction of a percent of the composition of the Earth. Silicon, just below carbon on the periodic table and one thousand times more abundant on Earth, has similar tastes in bonds, but has never been known to form the basis for life. Boron, another large part of Earth's crust, has bond properties that, like carbon and silicon, are very diverse but we know of no life-form based on boron.

That's because silicon- or boron-based chemistry cannot happen in a simi-lar way to organic (carbon) chemistry, and even if they could, live organisms couldn't form because the energy required to form and break their molecules would be unachievable. Carbon's small radius means it has relatively small atomic orbitals, and without the layers of atomic orbitals and consequent electronic shielding found in silicon or other elements, carbon can form very compact and energy-efficient bonds with nitrogen, with hydrogen, and even with itself. These molecules are the building blocks of proteins.

Carbon is a changeling: It can be bound to four different elements but be completely different from another carbon bound to the same four elements (chemists call this *chirality*). Carbon is promiscuous: It goes beyond its marriage to two or three other partners and brings in yet another. It may have very stable single, double, or triple bonds, while still having electrons left over for a new bond somewhere else. A dashing suitor named Hydrogen, perhaps. Carbon's relationship with hydrogen is especially romantic. Self-assembling structures that could be a precursor to some form of self-animating life are rarely found in nature, but carbon and hydrogen do this. Joined together they repel electrons (hydrophobicity) and so are often called fatty acids. Place some of these fatty acids in proximity and they self-assemble into proto-cells.

Mitochondria are organelles that produce energy for the body from the sugars made during photosynthesis. There can be tens of thousands of these in each plant cell. The energy they create is stored in adenosine triphosphate (ATP) molecules, the universal currency for biological energy. You can think of sugar as fuels, mitochondria as generators, and ATP as the battery. Hydrogen in the embrace of carbon is the proton-motive force that catabolizes ATP — carbon provides the wiring system that lets energy pulse. When we starve the soil of carbon, we starve ourselves. When we send it off to the atmosphere, we stitch our own death shroud.

Treated poorly, carbon can be a jealous and vengeful partner. Scorned or abused, it can become, quite suddenly, your worst nightmare. It holds this entire planet in its grip and with a small flick of its finger can bring human hubris to heel.

We need to get past our fear of this element. Embrace her, and make her your friend. She can as easily save you as destroy you. Wouldn't you rather she welcomed you back . . .

. . . to her garden planet?

What we have shown in this book is that carbon cash comes from carbon caching. The rewards are there for the taking. Let's take them.

ACKNOWLEDGMENTS

Without the inspiration, encouragement, and assistance we were provided this book would not have been possible. To these individuals, and any others whom we have inadvertently omitted, we express our profound gratitude. We hope this book returns a small amount to you for what you gave to us. Our thanks to Liora Adler, Paul Allen, Kevin Anderson, Mikoto Araki, Norman Baker, Margo Baldwin, Steve Barber, David Blume, Mary Ellen Bowen, Joe Brewer, George Brown, Sarah Brownell, Eoin Campbell, Jane Case, Bob Cimino, Raines Cohen, Suzanne Cooper, Pachi Coquette, Paddy Cosgrave, Annette Cowie, Brando Crespi, Ken Crouse, Jason Deptula, David Derbowka, Brian Dillon, Marty Dodge, Cassidy Draper, Gerald Dunst, Bjorn Embren, Dean Farago, Michael Farrelly, Christiana Figueres, Tim Flannery, Gloria Flora, Veronica Valenzuela Gibson, Peter Gilmer, Tatiana Ginsberg, Brianne Goodspeed, Al Gore, Thomas Goreau, Ian Graham, Marian Grebanier, Bronson Griscom, Gayla Groom, Mattias Gustofsson, Doug Guyer, Charles A. S. Hall, Jeff Hallowell, Maddy Harlan, Paul Hawken, Joseph Hazelbaker, Tim Hegberg, C. Dale Hendricks, Kathryn Hill, Peter Hirst, James Hollister, Martin Holsinger, Bob Holzapfel, Josiah Hunt, Ross Jackson, Haichao Johnson, Stephen Joseph, Kosha Joubert, Sandor Katz, Karen Kenney, Rola Khouri, Bobby Klein, Ava Klinger, Kent Klitgaard, Mark Krawczyk, Olga Kuchukov, Gregory Landua, Andrew Langford, Ronal and Gretchen Larson, Raleigh Latham, Johannes Lehmann, Jonah Levine, John D. Liu, Laura Look, Jan Lundberg, Rob Lyman, Carl Maida, Emiliano Maletta, Eugene Marner, Duncan Martin, Justin and Anne Martin, Matthew McClure, David McConville, Frank Michael, Thomas Miles, Dan Miner, Christopher and Celini Nesbitt, Bernd Neugebauer, Trevor Nielsson, Emer O'Siochru, Santiago Obarrio, Geoff Oelsner, Ave Oit, Pan Genxing, James Patterson, Mike Pearson, Stephen Peel, Maria Piza, Christian Pulver, Greg Ramsey, Permaculture Realized, Sai Bhaskar Reddy Nakka, Guy Renaud, Mary Robinson, Johan Rockstrom, Wesley Roe,

Hans-Peter Schmidt, Patricia Scotland, Michael Shafer, Hayley Joyell Smith, Andrezej Stefaniak, Will Steffen, Frank Strie, Carl Swanson, Jack Swindlehurst, Sandra J. Thomson, Tom Trabold, Leobardo Velazquez, Ingvar Villido, Daniel Christian Wahl, Wang Li, Rob Wheeler, Kathy Willowoode, Kelpie Wilson, Stuart Muir Wilson, Michael Wittman, David Yarrow, and Christine Young.

Carbon Cousins

The carbon materials of interest in these pages blur the lines of ancient and emergent human-made carbonaceous materials — from mineral coal to biochar, activated carbon, and graphite. But while all these materials can look remarkably similar, at least to the naked eye, there are many differences. They vary in what they are made from, how they are made, what they are used for, and what characteristics are most relevant for their intended end uses. They also differ widely in their impact on the planet, either feeding and sheltering us or suffocating and slaughtering us.

COAL

The most ancient of the carbon cousins is coal. Coal is a flammable black rock. It makes up about 40 percent of the world's CO_2 emissions from fuels. It is 65–95 percent carbon and also contains hydrogen, sulfur, oxygen, nitrogen, and trace elements. Coal became a sedimentary rock from peat undergoing the pressure of rock laid on top of former swamps. All coal is derived from plant material. The harder forms, such as anthracite, were transformed at higher temperature and pressure — lots of layers of rock, volcanism, and earthquakes. While coal use is declining, it is still used extensively to produce electricity, steel, and cement.

COKE

Coal can be pyrolyzed (heated to high temperature in the absence of oxygen) to produce coke, an almost pure form of carbon with high energy density (56 megajoules/kilogram). Because coke burns even hotter than coal, it is used for smelting and refining, to extract metals from their ores, and to forge and anneal high-tempered steel and alloys.

PEAT

Peat is juvenile coal. It is marshland vegetation — trees, grasses, fungi, as well as other types of organic remains, such as insect and animal remains — preserved by the absence of oxygen in submerged sediments. While it has fewer concentrated calories than coke or coal, it is still mined for cooking and domestic heating in places where trees are scarce, such as Ireland, Scotland, and Central Russia.

Most modern peat bogs formed in high latitudes after the retreat of the glaciers at the end of the last ice age some nine thousand years ago, but world-wide, peatlands have been forming for 360 million years. The world's estimated 3 million square kilometers (1.16 million square miles) of peatland, which represent about 3 percent of Earth's land and freshwater surface, are capable of storing some 2 trillion tons of CO_2 — equivalent to about a hundred years' worth of fossil fuel emissions. By comparison, that is about two-thirds of the amount CO_2 now in Earth's atmosphere, but safely locked up in peat. Or is it?

The burning of peat bogs in Indonesia, with their large and deep growths containing more than 50 gigatons of carbon, continues to spew CO_2. Fires in Kalimantan and East Sumatra have released up to 40 percent of the amount released by fossil fuels — by themselves greater than the carbon uptake ability of the world's biosphere. More than one hundred peat fires, now burning since 1997, each year ignite new forest fires. Impossible to extinguish, they could continue to burn until their peat supply is exhausted around 2040.

Global demand for tree farms and palm oil plantations has sped up peat-land conversion in Southeast Asia. In Finland, peat is unsustainably harvested and fed to hungry electric power stations. But burning peat to make energy and heat is not the only thing peat is used for. Peat's insulating properties make it useful to industry. It might be found in cables or bearings. It was tradition-ally stuffed in cracks under roofs or in walls in rural areas. Smoke from peat fires is used to dry malted barley that gives Scotch the distinctive smoky flavor we call "peatiness." It is also harvested for use in horticulture and agriculture.

Like coal and coke, tar sands, and oil shale, peat is a fossil carbon best left unburned. The governments of the world are slowly (too slowly) moving to make the mining of all fossil fuels illegal, but until then, we need to look to commercial replacements that do not destroy the atmosphere.

CHARCOAL

Traditionally the word *charcoal* has been used to refer to carbonized biomass predominantly used for heating or cooking. Many people have the association of a charcoal briquette, but the briquette was not invented until 1897 and charcoal traces back to prehistory. It was used to draw animals on the walls of caves. The Neolithic "Ice Man" discovered under a melting glacier in September 1991 in the Ötztal Alps (hence his nickname, Ötzi) on the border between Austria and Italy had sixty-one tattoos made of subcutaneous charcoal. It is thought these were pain-relief treatments similar to acupressure or acupuncture to soothe arthritis.

When burned, wood becomes brown at 220°C (428°F), a deep brown-black after some time at 280°C (536°F), and an easily powdered ash at 310°C (590°F). Charcoal made when wood is burned without oxygen turns brown, soft, and friable at 300°C (572°F) and readily inflames at 380°C (716°F). When made at higher temperatures it takes a hard and brittle form and does not fire until heated to about 700°C (1292°F).

Charcoal is most often made from wood or wood by-products and may include various types of binders to keep the briquettes neat and tidy. When Henry Ford first started mass-producing automobiles, he did not want to waste the sawdust and wood scraps from making the bodies and frame, so they were transformed into briquettes. Ford Charcoal went on to become the Kingsford Company.

Charcoal production aims for high yields and so is usually done at relatively low temperatures (less than 320°C/608°F). This carbonization process is sometimes referred to as torrefaction. Torrefaction, really just singeing wood, dewaters biomass to render it more flammable, with low smoke emissions.

Common charcoal is made from peat, coal, wood, coconut shell, or petroleum. Traditional charcoal is made directly from hardwood material and holds the shapes of its feedstock. Pelletized or briquetted charcoal is made with mills and presses, and shaped either before or after carbonization.

Briquettes are made by compressing charcoal, typically made from sawdust and other wood by-products, with a binder and other additives. The binder is usually starch, and the press produces briquettes that are pillow-shaped. Briquettes may also include brown coal or anthracite (as heat source), mineral carbon (as heat source), borax, sodium nitrate (ignition aid), limestone (ash-whitening agent), raw sawdust (ignition aid), and other additives.

Sawdust briquette charcoal is made by compressing sawdust without binders or additives. It is the preferred charcoal in Taiwan, Korea, Greece, and the Middle East. It has a round hole through the center, with a hexagonal intersection. It is used primarily for barbecue as it produces no odor, no smoke, little ash, high heat, and is long burning — exceeding four hours.

Laboratory charcoal is obtained from the carbonization of sugar and is usually purified by boiling with acids to remove any mineral matter or other contaminants. The pyrolysis process may finish in a current of chlorine gas to remove the last traces of hydrogen.

Activated charcoal is produced especially for medical and industrial uses by heating common charcoal in the presence of steam or other gases, which causes the charcoal to develop more pores and to cleanse existing pores of impurities. Although it can be made using renewable biomass, it is often made from fossil fuels such as coal, petroleum-derived pitch, and peat. Activated charcoal production is energy-intensive, involving multiple stages requiring high heat, and so it comes with a heavy carbon footprint. While there are many different grades of activated carbon, most are costly, so researchers are on the hunt for something cheaper, less energy-intense, and more renewable.

Extruded charcoal is made by extruding either raw ground wood (subsequently carbonized) or carbonized wood into logs without use of a binder. The heat and pressure of the extruding process holds the "logs" together.

Japanese charcoal has had wood vinegar (pyroligneous acid) extracted by distillation during the charcoal making; it therefore produces almost no smell or smoke when burned. The charcoal of Japan is classified into three types:

1. Black charcoal is used for heating and cooking.
2. White charcoal (*binchōtan*) is used for wind chimes and xylophones. It is made by steaming ubame oak at high temperatures (1000°C/1832°F) to produce numerous small pores and a hard finish that sounds like metal when struck. Its pore structure can adsorb chemical substances so in Japan bits may be added to rice during cooking to remove the chlorinated tap water taste. It is also placed in shoes or cabinets to adsorb odors. Currently there are many binchōtan-based consumer products on the market, including socks, shirts, shampoo, and cosmetic products.
3. *Ogatan* is made from hardened sawdust. It is most often used in Izakaya or Yakiniku restaurants.

Appendix A

CARBON BLACK

Carbon black is the name of a common black pigment used in inks and paints, traditionally produced from charring organic materials such as wood or bone. It appears black because it reflects little light in the visible part of the spectrum, with an albedo near zero. Possibly carcinogenic when inhaled, it comes from the incomplete combustion of heavy petroleum products, including coal tar, ethylene cracking, and vegetable oil. It is used as a reinforcing filler in tires and other rubber products, including engine mounts or elastomers for aircraft vibration control.

GRAPHITE

Graphite is an extremely stable crystalline allotrope of carbon, a semimetal either mined from natural formations or made synthetically. China earns $7 billion per year from its graphite mines while the United States makes more than a billion dollars' worth in laboratories every year. Graphite's unique electrical, acoustic, lubricant, and thermal properties make it a much sought-after industrial product, and it can be found in anything from satellites and nuclear reactors to fishing rods and pencils.

Carbon materials manufacturing may be done in many ways, from low tech to high tech, from primitive, polluting kilns, to more modern technologies with sophisticated emission controls. While the feedstock might be from biomass, and theoretically renewable, in some parts of the world production of charcoal has caused massive deforestation. Indeed, Europeans first turned to burning peat and coal in the Middle Ages when the overexploitation of forests for charcoal led to shortages. Today the production and use of most of these carbon products is exacerbating climate change. While some may claim that charcoal is carbon-neutral, this is not the case if forests are being clear-cut to produce it.

APPENDIX B

Carbonizers

Combustion can be rapid, explosive, or spontaneous. Rapid combustion burns quickly, gives off a lot of energy, and is how some of us cook. It is also how we make steam-driven electricity or warm our homes with fuels like wood, coal, or heating oil. Explosions are even faster, producing dangerous amounts of heat and sound. Tamed explosions are how we drive our cars. Carbonization uses heat, steam, or chemicals to do the same thing as fire, in a more controlled way. In addition to woody biomass, gasifiers can be used for other common carbon-rich materials, including shredded tires, pulverized plastics, or municipal landfill wastes, but we are cautious not to call these human-made materials biochar. Although they can be pyrolyzed, those feedstocks won't exhibit soil fertility properties and may be detrimental to soils and rivers.

Most pyrolysis and carbonization reactions occur between 350°C and 500°C (662°F and 932°F). Pyrolysis, the first stage, focuses on the chemical breakdowns of long-chain sugars that release smaller molecules as gases. Carbonization, the second stage, focuses on stabilizing remaining carbon atoms into solid structures. One can think of pyrolysis and carbonization as simultaneous physical–chemical processes, producing pyrolytic gases and charcoal.

Carbonization converts biomass into biochar (in its safe-for-soil form) or pyrolysates (in its unsafe-for-soil form). Very slow reactions between carbonized biomass and oxygen slowly degrade exposed graphene bonds over the course of thousands to millions of years.

There are various types of pyrolysis and gasification technologies and the end products — charcoal, bio-oils, syngas, wood vinegar, heat, electricity — vary depending on feedstocks, dwell times, temperatures, and the type of thermochemical conversion — torrefaction, slow, fast, flash, or microwave. The properties of the carbon products can be endlessly tweaked by preprocessing feedstock, modifying production, or postprocessing. The research coming out on different types of biochar for different end uses is ramping up quickly, with more peer-reviewed papers published every day — more than twenty-seven hundred in the first four months of 2018.

The technologies run from no-cost to multi-million-dollar methods, from ancient to emergent, stationary to mobile units and batch or continuous machines. Carbonization technology is no longer a constraint for scaling up biochar production, and new variations continue to emerge. Here are a few that are widely used.

MINIMUM COOL

TLUDs

Top-loading updraft stoves (TLUDs) have been around for a few decades. Originally designed as clean-burning, batch-style cookstoves, these days they perform many more functions — from making supper to charging cell phones.

Although there are variations, the basic setup is a barrel within a barrel with some holes near the bottom to allow for airflow, a reducer on top to slow that airflow, and a chimney over the reducer to take away smoke. A wide variety of biomass can be charred, and biochar yields are typically in the 20 percent range by volume.

The Warm Hearts Foundation, a charity working with rural farmers in Mae Chaem, Thailand, commissioned 250 locally made "Jolly Rogers," a type of TLUD using 55-gallon barrels. After training local farmers on how to use the J-Ros, as the Thais call them, Warm Hearts paid farmers to carbonize mountains of corncobs left at a de-kerneling facility.

Once farmers had completed the harvest season, entire families showed up to process the cobs for the subsidy. Each family tended up to ten kilns and made roughly 800 kilograms (1,760 pounds) of biochar every day.

Flame Curtain (Kon-Tiki, Trench, Pyramid)

Cool slaw began as a remedy for wild boars. In rural Japan, there is an area between the valley farms and the high mountain forests called satoyama, the zone where the land is held as commons — used for hunting, gathering bamboo, and cutting small trees for shiitake logs. For family farms, the satoyama region is actually quite important, because it provides fuel, fiber, furniture, and food that can't easily be grown on their small landholdings.

As the population shifted away from rural areas into cities, the satoyama regions fell into disuse. The bamboo canebrakes grew dense and impenetrable.

They became habitat for wild boars that ranged into the rice paddies at night, rooting and browsing the tender plants. With bamboo groves so dense, hunting and trapping the boars was futile. In recent years the spiral downward accelerated, with more farms lost, satoyama neglected, and boars growing in numbers.

In 2009, an experiment was begun by the Hozu farming cooperative in Kyoto Prefecture in partnership with Ritsumeikan University and the Kameoka City government. For the government, the goal was to reduce Kameoka City's carbon footprint. The university offered its expertise to bring those different stakeholder visions together. Together, the three partners started the Carbon Minus Project, with a plan to harvest satoyama bamboo and make biochar.

With 338 of 352 farming households in the Hozu hamlet joining in, they began to experiment with just one small plot of "cool vegetables." Using smokeless kilns shaped like huge woks, the bamboo was charred, ground up, composted, and the finished biochar applied to the experimental plot. The cabbages became "cool slaw," a sensation at the grocery.

Tinkerers from Oregon to Australia to Switzerland began to modify traditional Japanese cone kilns several years ago and eventually came up with what is now referred to as the "flame curtain." The burn starts at the top of a pile laid within a cone- or inverted-pyramid-shaped metal kiln. Layers of biomass are added as the first layers start to ash and the flame above uses up available oxygen, burning the volatile gases, so that the lower layers are protected from oxidizing and remain as solid carbon. The most popular model is the Kon-Tiki, now in more than seventy countries.

Flame curtain kilns can be built at almost no cost by simply digging conical holes or trenches into the earth, but are more often made of metal. Because they can be any shape, they can be tailored to the dimensions of the feed-stock — long and narrow for bamboo or cane stalks, short and wide for wood logs, or conical for rice husks and corncobs.

Burns can be quenched either from the top or from below. The effect of quenching from below is similar to what happens when activated carbon is steam-treated. The steam cleans residues from the pore spaces and makes the char less hydrophobic. The alkalinized quench water can also be captured and stored for later use as a soil conditioner, liming agent, or mild cleaning liquid.

Some quench with "iron water" in order to magnetize the biochar. Magnetized biochars are better for filtration and remediation of soils containing heavy metals. Iron water can be made simply using a 55-gallon drum filled with old metals and water.

Instead of a liquid quench, snuffing is popular in water constrained areas — smothering the flame with an organic material such as manure or clay.

MEDIUM COOL

Curtain Burners

An air curtain burner is a large, refractory-lined box equipped with a powerful blower designed to ash biomass without soot, but can also be operated to produce biochar. Air Burners, Inc., has been making these burners for more than twenty years.

The US Forest Service investigated curtain burners and in 2002 recommended their use for safely carbonizing forest waste with almost no emissions. During a fuel-reduction project on the Siskiyou National Forest in May 2016, production rate was 800 kilograms of biochar per hour from a single mobile burner. The burners minimize preprocessing, chipping, and transportation. While the output of biochar is low (2 percent by volume), it is possible to increase that by quenching where water is available.

Many biomass energy systems require significant preprocessing, sorting, grinding, and chipping, as well as a secondary fuel source, such as natural gas, to support combustion. Portable air burners eliminate most of this preprocessing and the need for secondary fuel. They are excellent options for managing enormous amounts of forestry slash (98 percent mass reduction), and can supply power in three ranges: 100 kW, 500 kW, and 1,000 kW (1 Megawatt).

All Power Labs

All Power Labs (APL) is a ten-year-old California company that builds small-scale gasifiers that make power while also making biochar. Their PP25 Power Pallet ($60,000) can run continuously for up to twelve hours and generates 25 kW. Larger 50 kW to 150 kW models are in limited production. In addition to hardwood, softwood, and nutshells, APL's machines have also been tested to work with low-moisture-content crop residues such as corncobs, coconut shells, and palm kernel shells.

Biochar Solutions Inc.

BSI is based in Carbondale, Colorado, and sells continuous feed reactors capable of handling up to 1 ton per hour of biomass with a maximum moisture content of 20 percent and maximum biomass size of 4 inches. Their reactors yield up to 300 pounds or 1.5 cubic yards of biochar per hour (a 15 percent yield). Biochar is produced in two distinct sizes – small chip and fine powder. BSI also offers materials and services such as blending, compounding, packaging, technical and aesthetic design, and certification including organic and biodynamic.

Biomass Controls
Connecticut-based Biomass Controls has focused on building portable, modular biogenic refineries that remove 100 percent of pathogens. The unit can process high-moisture-content feedstock such as biosolids and food-processing residues. BC offers turnkey technology solutions to improve sanitation where access to wastewater treatment is lacking. Their biogenic refinery produces both heat and biochar and additional capabilities for generating electricity are being developed.

Ökozentrum
The Swiss company Ökozentrum designs pyrolizers for different food-processing plants with a current focus on coffee pulp. Their equipment can convert 2,200 tons per year of moist or dry residues into approximately 300 tons of biochar, generating 50 kW of power and 200 kW of heat in the process. The heat generated is used to dry beans. Plants have been built in Vietnam, Peru, and Brazil, with others currently in construction for customers in Africa and Eastern Europe.

Organilock
Organilock (LEI Products) markets a range of automated biomass burners capable of producing between 300,000 and 1,250,000 BTUs of heat – enough to boil 350 gallons of water while making 7–20 gallons (1–3 cubic feet) of biochar per hour. Feedstocks must be well screened and fairly small (less than 2 inches), but the burners can accept feedstock with up to 40 percent moisture.

Proton Power
Proton Power, Inc., has developed a cellulose-to-hydrogen power (CHyP) system that produces inexpensive, hydrogen-rich gas on demand with biochar and water as co-products. The system can process woody biomass, energy crops, and preprocessed municipal waste with up to 45 percent moisture. The

syngas can be burned directly to generate electricity or as a substitute fuel for diesel generators.

Pyreg

For the past decade the German company Pyreg has specialized in compact, continuous-feed, slow pyrolysis technology. They have sold more than twenty units throughout Europe and recently entered the US marketplace. Their customers include municipal governments, compost companies, agricultural concerns, recycling centers, and wastewater treatment plants.

The Pyreg 500 system can handle both low- and high-moisture-content feedstocks including sewage sludge, livestock manure, digestate, crop residues (such as husks or straw), forestry and green waste (such as wood chips or sawdust), and food-processing waste (such as shells, pits, or okara). On average the machine can process 1,400 tons of organic material per year, making 300 tons of biochar and 150 kW of heat.

Spanner Re²

The Re² Biomass CHP generates electricity (9 kW–68 kW) and heat (22 kW–123 kW) using modular wood gasifiers coupled to combined heat and power generators. The company was established in 2004 as a renewable energies unit of the Spanner Group. Today it has more than one hundred employees in Niederbayern, Germany, and has installed more than seven hundred biomass-to-biochar plants globally.

Super Stone Clean

If you have had *Hakusekikan* gourmet beer in Japan, you may know it comes from Iwamoto Company, founded in the thirty-third year of the Meiji era (1900) as a stonemasonry business, explaining the name for their biochar retort. The president of the company, Tetsuomi Iwamoto, is an avid inventor who also created a "slow house" monocoque construction method, a way of processing food wastes with stone powder, and a radionuclide-adsorbing mineral fertilizer for soil decontamination near Fukushima.

Super Stone Clean systems combine pyrolysis with magnetic field technology for site-of-generation waste processing. The fuel for the process is the waste itself, up to 3 tons per day (125 kg/hr) of almost any type of biomass feedstock or municipal solid waste. Process parameters such as residence time and temperature are controlled in order to influence biochar yields and properties, and a patented emissions-reduction system ensures clean processing, nontoxic emissions, and compliance with worldwide environmental regulations. Their class Z series, which can arrive on a small truck, processes plastic diapers, sewage sludge, polymers, wood, Styrofoam, vinyl, cardboard, paper, rubber, tires, food waste, PVC, and more. The 800–1200°C (1472–2192°F) kiln not only completely disassembles polymers, but also decomposes dioxins and chlorinated hydrocarbons into an ash that can be processed to create cinder blocks.

SynCraft

Austria's SynCraft Engineering builds gasification plants that produce heat, electricity (200 kW–500 kW), and biochar from dry forest residues and sawmill by-products. SynCraft's fixed bed system can produce varying amounts of biochar based on market demand, with up to 4.7 cubic meters of biochar per day possible. They currently have several systems operating in Austria and are shipping multiple units to Japan.

THE BIG CHILL

Ankur Scientific

Located in Gujarat, India, Ankur Scientific has been working in the waste-to-energy field since 1986. It handles municipal solid waste, poultry litter, palm kernel shells and other ag wastes, sewage, tires, paper, and all manner of woody biomass. With some 250 employees, Ankur offers more than sixty different gasifiers for electrification and biogas.

Benenv

The Jiangsu Benenv (Benefit the Environment) Environmental Technologies Company has central offices in Kanazawa City, Japan, and its factory in Yixing, China. It produces devices primarily designed to carbonize municipal or agricultural sludge. It provides small-to-medium modular solutions that range from mobile sludge dewatering service vehicles to rotary dryers and carbonizers using their own gas to operate and reducing sludge volume more than 90 percent.

Biogen

Based in the Dominican Republic, Biogen produces 250 kW–20 MWe biomass gasifiers. From site design to support, Biogen is a complete solutions company currently installing 1,500 MWe per year.

Carbon Terra

A continuous-feed, industrial-scale gasifier known as the Schottdorf-Meiler is made by Carbon Terra in Germany. Complete with a burner, steam turbine, and power generator, this unit can process many types of organic material with moisture content up to 40 percent. A daily input of 6 metric tons yields an output of 2 metric tons biochar and 300 kW of power from wood-gas. Carbon Terra only uses waste organic material, and a single unit can produce up to 700 Mt of biochar per year.

CoalTec

CoalTec, which started making large-scale gasification technology about ten years ago, positions itself as an environmental solutions company. Their equipment is used mostly with high-moisture-content feedstocks such as livestock manure or wet distillers' grain. The heat generated is used to dry organic material to between 20 percent and 25 percent. Temperatures are generally high (>760°C/1400°F), with biochar production of up to 12 tons per day. At prices ranging from $2.7 to $3 million, these carbonizers are for operations that manage enormous amounts of organic material — for example a dairy operation with forty-five hundred cows or a hog operation handling 75 million gallons of slurry waste.

Etia/Biogreen

French engineering company Etia designs, manufactures, and implements technologies to process municipal waste, crop residues, animal manure, and biosolids. Their Biogreen pyrolysis technology has been sold to thirty locations since 2003. A continuous-feed pyrolysis unit is programmed for specific dwell times and heating temperatures up to 800°C (1472°F), allowing for the production of biochars with varying characteristics.

PRME

PRM Energy Systems of Hot Springs, Arkansas, makes KC-Reactor gasification technology, now in industrial applications on five continents, converting over 8.5 million tons of biomass and other waste fuels to clean energy annually. One of the first to enter the business, PRME has had carbonizers operating around the clock for over thirty-two years, with a range of units that can gasify from 30 to 2,000 tons of waste biomass per day.

Pyrocal

A ten-year-old Australian company, Pyrocal is a developer, designer, fabricator, and operator of continuous carbonization systems for residual biomass. Operating in six countries, their mission addresses carbon products, waste management, and waste-to-energy solutions. Their continuous autothermal carbonization systems convert 5–50 tons per day and can specialize in residues that are often problematic, such as manures, biosolids, barn litter, cotton gin trash, grape pomace, and other green wastes.

———

To be sure, each type of biomass requires different technologies and business models to make it financially viable, and ecological analysis is needed to optimize for ecosystem services, but once you open up the nonagricultural applications for carbonized wastes, the sources no longer need to be "food grade," the cost of conversion becomes affordable, and at that point, potential feedstocks are unlimited.

NOTES

INTRODUCTION

1. Timothy J. Garrett, "No Way Out? The Bind in Seeking Global Prosperity alongside Mitigated Climate Change," *Earth System Dynamics* 3, no. 1 (2012): 1–17.

2. Bruce H. Lipton, *The Biology of Belief 10th Anniversary Edition: Unleashing the Power of Consciousness, Matter & Miracles* (Carlsbad, CA: Hay House, 2016).

3. "List of Frequently Asked Questions," International Biochar Initiative, https://biochar-international .org/faqs.

4. A 2007 survey of trends in the atmospheric CO_2 budget showed that CO_2 emissions from fossil fuel burning and industrial processes have been accelerating at a global scale, with their growth rate increasing from 1.1% per year for 1990–1999 to >3% per year for 2000–2004. In 2005, some 7.9 gigatons of carbon (GtC) per year derived from fossil fuels and 1.5 GtC per year from land use changes with the former growing rapidly over recent years while the latter remains nearly steady. Michael Raupach et al., "Global and Regional Drivers of Accelerating CO_2 Emissions," *Proceedings of the National Academy of Sciences* 104, no. 24 (2007): 10288–10293.

PART I: CARBON CHANGE: FROM NEMESIS TO ALLY

1. John G. Neihardt, *Black Elk Speaks: Being the Life Story of a Holy Man of the Oglala Sioux* (Lincoln: University of Nebraska Press, 1961), chapter 17.

CHAPTER 1: CHARLES KEELING'S CURVE

1. "Mauna Loa Observatory," Earth System Research Laboratory Global Monitoring Division, National Oceanic and Atmospheric Administration, https://www.esrl.noaa.gov/gmd/obop/mlo.

2. Steve Dorst, *50 Years Ago, LBJ on Climate Change*, video, 00:28, February 2012, https://www .youtube.com/watch?v=So2_NxQLGxQ.

3. Nathaniel Rich, "Losing Earth: The Decade We Almost Stopped Climate Change," *New York Times Magazine*, August 1, 2018.

4. Mark Hertsgaard, "While Washington Slept," *Vanity Fair*, January 1, 2010, https://www.vanityfair .com/news/2006/05/warming200605.

5. Mark New, Diana Liverman, Heike Schroder, and Kevin Anderson, "Four Degrees and Beyond: The Potential for a Global Temperature Increase of Four Degrees and Its Implications." *Philosophical Transactions: Series A, Mathematical, Physical, and Engineering Sciences* 369 (2011): 6–19.

6. Patrick T. Brown and Ken Caldeira, "Greater Future Global Warming Inferred from Earth's Recent Energy Budget," *Nature* 552, no. 7683 (2017): 45, https://www.nature.com/articles/nature24672.

7. David Coady, Ian Parry, Louis Sears, and Baoping Shang, "How Large Are Global Fossil Fuel Subsidies?," *World Development* 91 (2017): 11–27, https://www.sciencedirect.com/science/article /pii/S0305750X16304867#!.

8. Global emissions rose to an all-time high of 35.9 $GtCO_2$ in 2017, a 2 percent increase over 2016.

9. E. Kriegler et al., "Pathways Limiting Warming to 1.5°C: A Tale of Turning Around in No Time?," *Philosophical Transactions: Series A, Mathematical, Physical, Engineering, and Science* 376, no. 2119 (2018).

CHAPTER 2: X-AXIS OVER Y

1. J. Rogelj et al., "Zero Emission Targets as Long-Term Global Goals for Climate Protection," *Environmental Research Letters* 10, no. 105007 (2015).

2. Olivier Boucher et al., "Rethinking Climate Engineering Categorization in the Context of Climate Change Mitigation and Adaptation," *Wiley Interdisciplinary Reviews: Climate Change* 5, no. 1 (2014): 23–35.

3. John G. Shepherd, *Geoengineering the Climate: Science, Governance and Uncertainty* (London: The Royal Society, 2009).

4. Bronson W. Griscom, Justin Adams, Peter W. Ellis, Richard A. Houghton, Guy Lomax, Daniela A. Miteva, William H. Schlesinger, et al., "Natural Climate Solutions," *Proceedings of the National Academy of Sciences* 114, no. 44 (2017): 11645–11650.

5. Smith et al., "Biophysical and Economic Limits to Negative CO_2 Emissions," *Nature Climate Change* 6 (2016): 42–50.

6. G. Robertson et al., "Greenhouse Gases in Intensive Agriculture: Contributions of Individual Gases to the Radiative Forcing of the Atmosphere," *Science* 289 (2000): 1922–1925; S. Brown et al., *Baseline Greenhouse Gas Emissions and Removals for Forest, Range, and Agricultural Lands in California* (Winrock International, for the California Energy Commission, 2004); C. Li et al., "Carbon Sequestration in Arable Soils Is Likely to Increase Nitrous Oxide Emissions, Offsetting Reductions in Climate Radiative Forcing," *Climatic Change* 72 (2005): 321–338.

7. European Academies Science Advisory Council, *Negative Emission Technologies: What Role in Meeting Paris Agreement Targets?* (Schweinfurt, Germany: German National Academy of Sciences Leopoldina, 2018).

8. Bhupinder Pal Singh, Blake J. Hatton, Balwant Singh, Annette L. Cowie, and Amrit Kathuria, "Influence of Biochars on Nitrous Oxide Emission and Nitrogen Leaching from Two Contrasting Soils," *Journal of Environmental Quality* 39, no. 4 (2010): 1224–1235.

9. Mark G. Lawrence et al., "Evaluating Climate Geoengineering Proposals in the Context of the Paris Agreement Temperature Goals," *Nature Communications* 9, no. 1 (2018): 3734.

10. Johannes Lehmann and Stephen Joseph, eds., *Biochar for Environmental Management* (Amsterdam: Elsevier, 2008).

11. Rattan Lal, "Carbon Management in Agricultural Soils," *Mitigation and Adaptation Strategies for Global Change* 12 (2007): 303–322.

12. Pete Smith et al., "Agriculture," in *Climate Change 2007: Mitigation. Contribution of Working Group III to the Fourth Assessment Report of the Intergovernmental Panel on Climate Change* (Cambridge, UK: Cambridge University Press, 2007).

13. Smith et al., "Biophysical and Economic Limits to Negative CO_2 Emissions," *Nature Climate Change* 6 (2016): 42–50.

14. Robert Watson et al., eds., *Special Report: Land Use, Land-Use Change, and Forestry* (Cambridge, UK: Cambridge University Press, 2000).

15. Paul Hawken, ed., *Drawdown: The Most Comprehensive Plan Ever Proposed to Roll Back Global Warming* (New York: Penguin, 2017).

16. Klaus Lackner et al., "Carbon Dioxide Disposal in Carbonate Minerals," *Energy* 20 (1995): 1153–1170; Klaus Lackner, "A Guide to CO_2 Sequestration," *Science* 300 (2003): 1677–1678.

17. R. Schuiling and P. Krijgsman, "Enhanced Weathering: An Effective and Cheap Tool to Sequester CO_2," *Climatic Change* 74 (2006): 349–354; R. Schuiling and P. de Boer, "Rolling Stones: Fast Weathering of Olivine in Shallow Seas for Cost-Effective CO_2 Capture and Mitigation of Global Warming and Ocean Acidification," *Earth System Dynamics Discussions* 2 (2011): 551–568.

18. P. Köhler et al., "Geoengineering Potential of Artificially Enhanced Silicate Weathering of Olivine," *Proceedings of the National Academy of Sciences of the USA* 107 (2010): 20228–20233; P. Köhler et al., "Geoengineering Impact of Open Ocean Dissolution of Olivine on Atmospheric CO_2, Surface Ocean pH and Marine Biology," *Environmental Research Letters* 8 (2013): 014009.

19. Jens Hartmann et al., "Enhanced Chemical Weathering as a Geoengineering Strategy to Reduce Atmospheric Carbon Dioxide, Supply Nutrients, and Mitigate Ocean Acidification," *Reviews of Geophysics* 51, no. 2 (2013): 113–149.

20. Hawken, note 11.

21. "Marine Permaculture," Climate Foundation, http://www.climatefoundation.org/marine -permaculture.html.

22. E. Sanz-Perez et al., "Direct Air Capture of CO_2 from Ambient Air," *Chemical Reviews* 116 (2016): 11840–11876.

23. Allen B. Wright, Klaus S. Lackner, and Ursula Ginster, "Method and Apparatus for Extracting Carbon Dioxide from Air," US Patent 9,266,052, issued February 23, 2016.

24. Carnegie Council for Ethics in International Affairs, *Phil Renforth: Carbon Dioxide Removal (CDR) and Ocean Alkalinity*, video, 10:09, November 20, 2017, https://www.youtube.com/watch? v=J6uQj2t34HU.

25. Ejeong Baik et al., "Geospatial Analysis of Near-Term Potential for Carbon-Negative Bioenergy in the United States," *Proceedings of the National Academy of Sciences* 115, no. 13 (2018): 3290–3295.

26. James J. Dooley, "Estimating the Supply and Demand for Deep Geologic CO2 Storage Capacity over the Course of the 21st Century: A Meta-Analysis of the Literature," *Energy Procedia* 37 (2013): 5141–5150.

27. Dominic Woolf, Johannes Lehmann, and David R. Lee, "Optimal Bioenergy Power Generation for Climate Change Mitigation with or without Carbon Sequestration," *Nature Communications* 7 (2016): 13160.

28. Liam Brennan and Philip Owende, "Biofuels from Microalgae — a Review of Technologies for Production, Processing, and Extractions of Biofuels and Co-Products," *Renewable and Sustainable Energy Reviews* 14, no. 2 (2010): 557–577.

29. Tara Shirvani, Xiaoyu Yan, Oliver R. Inderwildi, Peter P. Edwards, and David A. King, "Life Cycle Energy and Greenhouse Gas Analysis for Algae-Derived Biodiesel," *Energy & Environmental Science* 4, no. 10 (2011): 3773–3778.

30. M. H. Langholtz, B. J. Stokes, and L. M. Eaton, *2016 Billion-Ton Report* (Washington, DC: Office of Energy Efficiency & Renewable Energy, 2016).

CHAPTER 3: GOD'S OWN ATOM

1. The triple-alpha condition occurs between about 7.3 and 7.9 MeV to produce sufficient carbon for life to exist, and the solar reaction must be further "fine-tuned" to between 7.596 MeV and 7.716 MeV to produce the abundant level of carbon-12 observed in nature.

2. David Yarrow, *Let Freedom Ring: Carbon, Minerals & Microbes* (Turtle Island Sanctuary, 2013).

3. Hugh McLaughlin et al., "All Biochars Are Not Created Equal, and How to Tell Them Apart," presentation, North American Biochar Conference, Boulder, CO, October 2009, http://www .biochar-international.org/node/1029.

CHAPTER 4: CARBON CASCADES

1. William McDonough and Michael Braungart, *Cradle to Cradle: Remaking the Way We Make Things* (North Point Press, 2010).

2. Daniel C. Wahl, *Designing Regenerative Cultures* (Triarchy Press via PublishDrive, 2016).

3. Cristina-Maria Iordan et al., "Contribution of Harvested Wood Products to Negative Emissions: Historical Trends in Norway, Sweden and Finland and Future Projections under the Shared Socioeconomic Pathways," presentation, International Conference on Negative CO_2 Emissions, Gothenburg, Sweden, May 23, 2018.

4. Johannes Lehmann, "Bio-Energy in the Black," *Frontiers in Ecology and the Environment* 5, no. 7 (2007): 381–387.

5. Gary Schnitkey, "2017 Crop Budgets, 2016 Crop Returns, and 2016 Incomes," *Farmdoc Daily* 7 (2017): 134, http://farmdocdaily.illinois.edu/2016/09/2017-crop-budgets-2016-crop-returns-incomes.html.

6. Dominic Woolf et al., "Sustainable Biochar to Mitigate Global Climate Change," *Nature Communications* 1 (2010): 56, https://www.nature.com/articles/ncomms1053?page=20.

7. Kate Raworth, *Doughnut Economics: Seven Ways to Think Like a 21st-Century Economist* (White River Junction, VT: Chelsea Green Publishing, 2017).

8. Kolby W. Smith, Maosheng Zhao, and Steven W. Running, "Global Bioenergy Capacity as Constrained by Observed Biospheric Productivity Rates," *BioScience* 62, no. 10 (2012): 911–922, https://doi.org/10.1525/bio.2012.62.10.11; Richard Heinberg, *Afterburn: Society Beyond Fossil Fuels* (Gabriola Island, Canada: New Society Publishers, 2015); "U.S. SHALE OIL PRODUCTION UPDATE: Financial Carnage Continues to Gut Industry," *SRSRocco Report*, November 14, 2017, https://srsroccoreport.com/u-s-shale-oil-production-update-financial-carnage-continues-to-gut-industry.

CHAPTER 5: CARBON HARDSCAPING

1. Jorge de Brito and Rui Silva, "Current Status on the Use of Recycled Aggregates in Concrete: Where Do We Go from Here?," *RILEM Technical Letters* 1 (2016): 1–5.

2. Sonia Paul, "How India's 'Sand Mafia' Pillages Land, Terrorizes People, and Gets Away with It," *Vice News*, October 7, 2015, https://news.vice.com/article/how-indias-sand-mafia-pillages-land-terrorizes-people-and-gets-away-with-it.

3. Jakob Villioth, "Building an Economy on Quicksand," Environmental Justice Organizations, Liabilities and Trade, August 5, 2014, http://www.ejolt.org/2014/08/building-an-economy-on-quicksand.

4. According to a 2013 survey by IBI, the average price for biochar in the US was $1.29 per pound or $2,580 per ton, though prices have started to come down as more production capacities comes online.

5. Won Chang Choi, Hyun Do Yun, and Jae Yeon Lee, "Mechanical Properties of Mortar Containing Bio-Char from Pyrolysis," *Journal of the Korea Institute for Structural Maintenance and Inspection* 16, no. 3 (2012): 67–74.

6. Ismael Justo-Reinoso et al., "Fine Aggregate Substitution by Granular Activated Carbon Can Improve Physical and Mechanical Properties of Cement Mortars," *Construction and Building Materials* 164 (2018): 750–759.

7. Hans-Peter Schmidt, "The Use of Biochar as Building Material – Cities as Carbon Sinks," *The Biochar Journal* (2014).

8. Hans-Peter Schmidt, personal communication to authors, June 1, 2018.

9. Souradeep Gupta, Harn Wei Kua, and Sze Dai Pang, "Healing Cement Mortar by Immobilization of Bacteria in Biochar: An Integrated Approach of Self-Healing and Carbon Sequestration," *Cement and Concrete Composites* 86 (2018): 238–254.

10. Khalifa S. Al-Jabri et al., "Concrete Blocks for Thermal Insulation in Hot Climate," *Cement and Concrete Research* 35, no. 8 (2005): 1472–1479.

11. Christopher Alexander et al., *A Pattern Language* (Oxford University Press, 1977), p. 958.

12. Ali Akhtar and Ajit K. Sarmah, "Novel Biochar-Concrete Composites: Manufacturing, Characterization and Evaluation of the Mechanical Properties," *Science of the Total Environment* 616 (2018): 408–416.

13. Luciana Restuccia and Giuseppe Andrea Ferro, "Promising Low Cost Carbon-Based Materials to Improve Strength and Toughness in Cement Composites," *Construction and Building Materials* 126 (2016): 1034–1043.

14. Won Chang Choi et al., "Mechanical Properties of Mortar Containing Bio-Char from Pyrolysis."

15. Emad Benhelal et al., "Global Strategies and Potentials to Curb CO_2 Emissions in Cement Industry," *Journal of Cleaner Production* 51 (2013): 142–161.

16. *The Asphalt Paving Industry: A Global Perspective,* 2nd ed. (Brussels, Belgium, and Lanham, MD: European Asphalt Pavement Association and National Asphalt Pavement Association, 2011).

17. Siti Nur Amiera Jeffry et al., "Effects of Nanocharcoal Coconut-Shell Ash on the Physical and Rheological Properties of Bitumen," *Construction and Building Materials* 158 (2018): 1–10.

18. Sheng Zhao et al., "Laboratory Investigation of Biochar-Modified Asphalt Mixture," *Transportation Research Record: Journal of the Transportation Research Board* 2445 (2014): 56–63.

19. *The Bitumen Industry A Global Perspective,* 3rd ed. (Lexington, KY: The Asphalt Institute, 2015).

20. Noor Azah Abdul Ramana et al., "A Review on the Application of Bio-Oil as an Additive for Asphalt," *Jurnal Teknologi* 72, no. 5 (2015).

CHAPTER 6: FANTASTIC PLASTIC

1. Jinhong Lü et al., "Use of Rice Straw Biochar Simultaneously as the Sustained Release Carrier of Herbicides and Soil Amendment for Their Reduced Leaching," *Journal of Agricultural and Food Chemistry* 60, no. 26 (2012): 6463–6470, https://pubs.acs.org/doi/abs/10.1021/jf3009734.

2. Lauren Hale, Madeline Luth, and David Crowley, "Biochar Characteristics Relate to Its Utility as an Alternative Soil Inoculum Carrier to Peat and Vermiculite," *Soil Biology and Biochemistry* 81 (2015): 228–235.

3. Aamer Khan et al., "Low-Cost Carbon Fillers to Improve Mechanical Properties and Conductivity of Epoxy Composites," *Polymers* 9, no. 12 (2017): 642.

4. Tim Flannery, *Atmosphere of Hope: Searching for Solutions to the Climate Crisis* (New York: Atlantic Monthly Press, 2014).

5. G. Richard and James E. Johnson, "Measuring Standing Trees and Logs," Virginia Cooperative Extension, http://pubs.ext.vt.edu/420/420-560/420-560.html.

6. "Wood Vinegar Uses," VerdiLife, https://www.verdilife.com/uses.

7. TMR Research, "Wood Vinegar Market — Global Industry Analysis, Size, Share, Growth, Trends and Forecast 2017–2025," September 2018.

8. David DeVallance, Gloria Oporto, and Patrick Quigley, "Hardwood Biochar as a Replacement for Wood Flour in Wood-Polypropylene Composites," *Journal of Elastomers and Plastics* 48, no. 6 (2015): 510–522.

CHAPTER 7: PAPER CHASE

1. "Green Packaging Market (Reusable Packaging, Recycled Content Packaging and Degradable Packaging) for Personal Care, Food & Beverages, Healthcare and Other Applications: Global Industry Perspective, Comprehensive Analysis, Size, Share, Growth, Segment, Trends and Forecast, 2015–2021," Zion Market Research, September 14, 2016, https://www.zionmarket research.com/report/green-packaging-market.

2. Krishna R. Reddy et al., "Enhanced Microbial Methane Oxidation in Landfill Cover Soil Amended with Biochar," *Journal of Geotechnical and Geoenvironmental Engineering* 140, no. 9 (2014): 04014047.

3. Meha Rungta, "Carbon Molecular Sieve Dense Film Membranes for Ethylene/Ethane Separations," PhD diss., Georgia Institute of Technology, 2012.

4. Suiyi Li, Xiaoyan Li, Chuchu Chen, Haiying Wang, Qiaoyun Deng, Meng Gong, and Dagang Li, "Development of Electrically Conductive Nano Bamboo Charcoal/Ultra-High Molecular

Weight Polyethylene Composites with a Segregated Network," *Composites Science and Technology* 132 (2016): 31–37, https://www.sciencedirect.com/science/article/pii/S0266353816305681.

5. See our Instagram video here: https://www.instagram.com/p/BVU8UCjhfc5.

6. "A Future-Oriented Paper Mill," Vicat, http://www.vicat.com/Activities/Other-products-and -services/Paper.

7. Jarkko Gronfors, "Use of Fillers in Paper and Paperboard Grades," final thesis, Tampere University of Applied Sciences, 2010.

CHAPTER 8: FILTRATION NATION

1. Arief Ismail et al., "Evaluation of Biochar Soil Amendments in Reducing Soil and Water Pollution from Total and Fecal Coliforms in Poultry Manure," *Canadian Biosystems Engineering* 58 (2016), http://www.csbe-scgab.ca/docs/journal/58/C16286.pdf.

2. Sanjay K. Mohanty et al., "Efficacy of Biochar to Remove *Escherichia coli* from Stormwater under Steady and Intermittent Flow," *Water Research* 61 (2014): 288–296.

3. Michael Kimmelman, "Jakarta Is Sinking So Fast, It Could End Up Underwater," *New York Times*, December 21; 2017, https://www.nytimes.com/interactive/2017/12/21/world/asia/jakarta -sinking-climate.html.

4. S. Dillon, "Mexico City Journal; Capital's Downfall Caused by Drinking . . . of Water," *New York Times*, January 29, 1998, http://www.nytimes.com/1998/01/29/world/mexico-city-journal -capital-s-downfall-caused-by-drinking-of-water.html.

5. "Drinking Water Fact Sheet," World Health Organization, February 7, 2018, http://www.who.int /mediacentre/factsheets/fs391/en.

6. Suşan Cosier, "Who Pays for Mine Cleanup after Big Coal Goes Bankrupt?," National Resources Defence Council, November 17, 2014, https://www.nrdc.org/onearth/ who-pays-mine-cleanup-after-big-coal-goes-bankrupt.

7. Susan Cosier, "This Land Is Mine Land — Abandoned Miles Are Polluting Western Waterways," National Resources Defence Council, November 17, 2014, https://www.nrdc.org/onearth/land -mine-land.

8. Luke Beesley and Marta Marmiroli, "The Immobilisation and Retention of Soluble Arsenic, Cadmium and Zinc by Biochar," *Environmental Pollution* 159, no. 2 (2011): 474–480.

9. Luke Beesley et al., "Biochar Addition to an Arsenic Contaminated Soil Increases Arsenic Concentrations in the Pore Water but Reduces Uptake to Tomato Plants (*Solanum lycopersicum* L.)," *Science of the Total Environment* 454 (2013): 598–603.

10. Luke Beesley, Eduardo Moreno-Jiménez, and Jose L. Gomez-Eyles, "Effects of Biochar and Greenwaste Compost Amendments on Mobility, Bioavailability and Toxicity of Inorganic and Organic Contaminants in a Multi-Element Polluted Soil," *Environmental Pollution* 158, no. 6 (2010): 2282–2287.

11. Luke Beesley, Eduardo Moreno-Jiménez, and Jose L. Gomez-Eyles, "Effects of Biochar and Greenwaste Compost Amendments."

12. Peng Liu, "Stabilization of Mercury in River Water and Sediment Using Biochars," doctoral thesis, University of Waterloo, 2016.

13. Sara Fellin, "Mercury Stabilization in Contaminated Sediment by Co-Blending with Solid-Phase Reactive Media," master's thesis, University of Waterloo, 2016.

14. Katerina Břendová et al., "Utilization of Biochar and Activated Carbon to Reduce Cd, Pb and Zn Phytoavailability and Phytotoxicity for Plants," *Journal of Environmental Management* 181 (2016): 637–645.

15. Edward A. Martell, "Tobacco Radioactivity and Cancer in Smokers: Alpha Interactions with Chromosomes of Cells Surrounding Insoluble Radioactive Smoke Particles May Cause Cancer

and Contribute to Early Atherosclerosis Development in Cigarette Smokers," *American Scientist* 63, no. 4 (1975): 404–412.

16. Nadia Karami et al., "Efficiency of Green Waste Compost and Biochar Soil Amendments for Reducing Lead and Copper Mobility and Uptake to Ryegrass," *Journal of Hazardous Materials* 191, no. 1–3 (2011): 41–48.

17. Genxing Pan, David Crowley, and Johannes Lehmann, *Burn to Air or Burial in Soil: The Fate of China's Straw Residues* (Washington, DC: International Biochar Initiative, 2011).

CHAPTER 9: CATTLE CARBS AND LEAFY GREENBACKS

1. D. Reddy et al., "Valorisation of Post-Sorption Materials: Opportunities, Strategies, and Challenges," *Advances in Colloid and Interface Science* 242 (2017): 35–58.

2. Brent Kim et al., *Industrial Food Animal Production — Examining the Impact of the Pew Commission's Priority Recommendations* (Baltimore, MD: John Hopkins Center for a Livable Future, 2013), https://www.jhsph.edu/research/centers-and-institutes/johns-hopkins-center-for-a-livable -future/_pdf/research/clf_reports/CLF-PEW-for%20Web.pdf.

3. PR Newswire, "Animal Health Market Size, Share & Trends Analysis Report by Animal Type," August 10, 2018.

4. S. B. Nageswara Rao and R. C. Chopra, "Influence of Sodium Bentonite and Activated Charcoal on Aflatoxin M1 Excretion in Milk of Goats," *Small Ruminant Research* 41, no. 3, (2001): 203–213, https://doi.org/10.1016/S0921-4488(01)00216-4.

5. Desert Hope, "Who Are the Players in the Pharmaceutical Industry (Big Pharma)?" https:// deserthopetreatment.com/big-pharma.

6. Stephen Joseph et al., "Feeding Biochar to Cows: An Innovative Solution for Improving Soil Fertility and Farm Productivity," *Pedosphere* 25, no. 5 (2015).

7. Lauren Celenza, *Whole-Farm Biochar System Boosts Productivity, Stores Carbon, Cuts Inputs and Emissions* (Floreat, Australia: Western Australia No-Tillage Farmers Association, 2016), http:// www.wantfa.com.au/wp-content/uploads/2016/04/WANTFA_NF_Summer2015_Biochar -002.pdf.

8. Nancy Glazier, *Gypsum Bedding in Long-Term Manure Storage May Create Dangerous Conditions* (Ithaca, NY: Cornell Cooperative Extension, 2013), https://scnydfc.cce.cornell.edu /submission.php?id=303.

9. Robb Meinen, "Gypsum Bedding — Risks and Recommendations for Manure Handling," presentation, LPELC Waste to Worth Conference, Denver, CO, April 4, 2013.

10. Michael DuPont, *Concepts of Innoculated Deep Litter Systems — Waste Management for Small Scale Piggeries*, video, 13:16, May 26, 2017, https://www.youtube.com/watch?v=d-5tkCgrwUA.

11. George Ford, "Iowa's Average Corn Yield Rose in 2013, Soybean Yield Unchanged," *The Gazette* (Cedar Rapids, IA), February 14, 2014, http://www.thegazette.com/2014/02/21/iowas -average-corn-yield-rose-in-2013-soybean-yield-unchanged.

12. Claire Dibble, "Acres per Finished Steer of Feedlot vs. Grassfed Beef," clairedibble.com, July 9, 2012, http://www.claredibble.com/2012/07/09/acre-per-pound-of-feedlot-vs-grassfed-beef; Gaston Neffen, "Un Feedlot que Produce Carne 'a Medida' y Ahora También Energía," *Clarín Rural* (Argentina), February 27, 2018, https://www.clarin.com/rural/feedlot-produce-carne -medida-ahora-energia_0_HkmSclzdG.html.

13. M. H. Langholtz, B. J. Stokes, and L. M. Eaton (leads), "2016 Billion-Ton Report: Advancing Domestic Resources for a Thriving Bioeconomy, Volume 1: Economic Availability of Feedstocks. ORNL/TM-2016/160," US Department of Energy, Oak Ridge National Laboratory, Oak Ridge, TN, 2016, doi:10.2172/1271651. http://energy.gov/eere/bioenergy /2016-billion-ton-report.

14. Yessie W. Sari, Utami Syafitri, Johan P. M. Sanders, and Marieke E. Bruins, "How Biomass Composition Determines Protein Extractability," *Industrial Crops and Products* 70 (2015): 125–133, http://dx.doi.org/10.1016/j.indcrop.2015.03.020.

15. Devin Takara and Samir Kumar Khanal, "Green Processing of Tropical Banagrass into Biofuel and Biobased Products: An Innovative Biorefinery Approach," *Bioresource Technology* 102, no. 2 (2011): 1587–1592.

CHAPTER 10: MANURE HAPPENS

1. The average world price for unblended biochar at factory at this writing is $175–$250 per ton (http://www.alibaba.com, retrieved May 26, 2018). Typical values for biochar dry bulk densities range 5–20 pounds per cubic foot, or 135–540 pounds per cubic yard, or 80–320 kilograms per cubic meter.

2. Oliver Milman, "'On a Hot Day, It's Horrific': Alabama Kicks Up a Stink Over Shipments of New York Poo," *The Guardian*, March 11, 2018, https://www.theguardian.com/us-news/2018/mar/11/alabama-new-york-poo-sewage-environment.

3. Jaio Cai, "Effects and Optimization of the Use of Biochar in Anaerobic Digestion of Food Wastes," *Waste Management and Research* 35, no. 5 (2016): 409–416, http://journals.sagepub.com/doi/abs/10.1177/0734242X16634196.

4. Brian Dougherty et al., "Can Biochar Covers Reduce Emissions from Manure Lagoons While Capturing Nutrients?," *Journal of Environmental Quality* 46, no. 3 (2017): 659–666.

5. Simon Kizito et al., "Evaluation of Slow Pyrolyzed Wood and Rice Husks Biochar for Adsorption of Ammonium Nitrogen from Piggery Manure Anaerobic Digestate Slurry," *Science of the Total Environment*, 505 (2015): 102–112.

6. Michael A. Holly, *Abatement of Greenhouse Gas and Ammonia Emissions from Storage and Land Application of Dairy Manure* (Madison: University of Wisconsin, 2016).

7. Miguel Sanchez-Monedero et al., "Role of Biochar as an Additive in Organic Waste Composting," *Bioresource Technology* 247 (2018): 1155–1164.

8. John Ross, "Fate of Micropollutants During Pyrolysis of Biosolids," master's dissertation, Marquette University, 2014, https://epublications.marquette.edu/cgi/viewcontent.cgi?referer=&httpsredir=1&article=1283&context=theses_open.

9. T. C. Hoffman, D. H. Zitomer, and Patrick J. McNamara, "Pyrolysis of Wastewater Biosolids Significantly Reduces Estrogenicity," *Journal of Hazardous Materials* 317 (2016): 579–584.

10. Alexandros Kelessidis and Athanasios S. Stasinakis, "Comparative Study of the Methods Used for Treatment and Final Disposal of Sewage Sludge in European Countries," *Waste Management* 32, no. 6 (2012): 1186–1195.

11. A. Castri et al., "The Use of Water in the Incineration Plant of Torino," presentation, Eleventh International Waste Management and Landfill Symposium, Cagliari, Italy, October 1–5, 2007, http://www.limpezapublica.com.br/textos/122.pdf.

12. T. R. Bridle and D. Pritchard, "Energy and Nutrient Recovery from Sewage Sludge via Pyrolysis," *Water Science and Technology* 50, no. 9 (2004): 169–175.

13. Marta Camps-Arbestain et al., "A Biochar Classification System and Associated Test Methods," *Biochar for Environmental Management: Science, Technology and Implementation* (2015): 165–193.

14. Amanda Saint, "Biocoal from Organic Waste," *Eniday*, https://www.eniday.com/en/education_en/biocoal-from-organic-waste.

15. Martina Flörke et al., "Domestic and Industrial Water Uses of the Past 60 Years as a Mirror of Socio-Economic Development: A Global Simulation Study," *Global Environmental Change* 23, no. 1 (2013): 144–156.

16. C. Colose, "Is Methane a 'Better' Greenhouse Gas than CO_2?," *Climate Change: An Analysis of Key Questions* (blog), November 10, 2008, https://chriscolose.wordpress.com/2008/11/10/methane

Notes

-and-co2; David Archer, "Methane Hydrate Stability and Anthropogenic Climate Change," *Biogeosciences Discussions* 4, no. 2 (2007): 993–1057.

CHAPTER 11: BLUE-GREEN REVOLUTION

1. Amy T. Hansen et al., "Contribution of Wetlands to Nitrate Removal at the Watershed Scale," *Nature Geoscience* 11, no. 2 (2018): 127.

2. Loretto Contreras-Porcia et al., "Biochar Production from Seaweeds," *Protocols for Macroalgae Research* (2018): 175.

3. Tim R. Searchinger et al., "Do Biofuel Policies Seek to Cut Emissions by Cutting Food?," *Science* 347, no. 6229 (2015): 1420–1422; Pete Smith et al., "Biophysical and Economic Limits to Negative CO_2 Emissions," *Nature Climate Change* 6, no. 1 (2016): 42.

4. Adam D. Hughes et al., "Does Seaweed Offer a Solution for Bioenergy with Biological Carbon Capture and Storage?," *Greenhouse Gases: Science and Technology* 2, no. 6 (2012): 402–407.

5. Michael I. Bird et al., "Algal Biochar — Production and Properties," *Bioresource Technology* 102, no. 2 (2011): 1886–1891; David A. Roberts et al., "Biochar from Commercially Cultivated Seaweed for Soil Amelioration," *Scientific Reports* 5 (2015): 9665.

6. Michael I. Bird et al., "Algal Biochar: Effects and Applications," *GCB Bioenergy* 4, no. 1 (2012): 61–69.

7. Charles H. Greene et al., "Marine Microalgae: Climate, Energy, and Food Security from the Sea," *Oceanography* 10 (2015): 249–265.

8. Charles H. Greene et al., "Geoengineering, Marine Microalgae, and Climate Stabilization in the 21st Century," *Earth's Future* 5, no. 3 (2017): 278–284.

9. Christina E. Canter et al., "Implications of Widespread Algal Biofuels Production on Macronutrient Fertilizer Supplies: Nutrient Demand and Evaluation of Potential Alternate Nutrient Sources," *Applied Energy* 143 (2015): 71–80.

10. Brian J. Walsh et al., "New Feed Sources Key to Ambitious Climate Targets," *Carbon Balance and Management* 10, no. 1 (2015): 26; M. J. Walsh et al., "Large-Scale Emissions, Land-Use and Water Savings through Algal Food and Fuel Co-Production," *Environmental Research Letters* 11 (2016): 114006.

11. Mark Ashton Zeller et al., "Bioplastics and Their Thermoplastic Blends from Spirulina and Chlorella Microalgae," *Journal of Applied Polymer Science* 130, no. 5 (2013): 3263–3275; Alexander Otto et al., "Closing the Loop: Captured CO_2 as a Feedstock in the Chemical Industry," *Energy & Environmental Science* 8, no. 11 (2015): 3283–3297.

12. Caroline A. Ochieng and Paul L. A. Erftemeijer, "Accumulation of Seagrass Beach Cast along the Kenyan Coast: A Quantitative Assessment," *Aquatic Botany* 65, no. 1–4 (1999): 221–238. See also: Anthony W. D. Larkum, Robert J. Orth, and Carlos M. Duarte, eds., *Seagrasses: Biology, Ecology and Conservation* (Netherlands: Springer, 2007).

13. "Tiempo: Reunión emergente," *Por Esto!*, March 25, 2018.

14. Antoine de Ramon et al., "Negative Carbon via Ocean Afforestation," *Process Safety and Environmental Protection* 90, no. 6 (2012): 467–474.

15. Information from http://www.alibaba.com, retrieved May 26, 2018.

16. Hans-Peter Schmidt et al., "Forest Gardens for Closing the Global Carbon Cycle," *The Biochar Journal* (2017).

17. Johannes Lehmann and Stephen Joseph, eds., *Biochar for Environmental Management* (Amsterdam: Elsevier, 2008).

18. Marjorie Kauffmann et al., "Evidence for Palaeo-Wildfire in the Late Permian Palaeotropics — Charcoal from the Motuca Formation in the Parnaíba Basin, Brazil," *Palaeogeography, Palaeoclimatology, Palaeoecology* 450 (2016): 122–128.

19. Jaimie Fullerton, "China to Plant Forest the Size of Ireland in Bid to Become World Leader in Conservation," *The Telegraph*, January 5, 2018, https://www.telegraph.co.uk/news/2018/01/05/china-plant-forest-size-ireland-bid-become-world-leader-conservation.
20. Coppice Agroforestry — Perennial Silviculture for the 21st Century, http://www.coppiceagro forestry.com.
21. Davina Van Goethem et al., "Correction: Seasonal, Diurnal and Vertical Variation of Chlorophyll Fluorescence on *Phyllostachys humilis* in Ireland," *PLoS One* 8, no. 11 (2013).
22. Inspire Cuba, http://www.inspirecuba.org.
23. Eric M. Karp et al., "Renewable Acrylonitrile Production," *Science* 358, no. 6368 (2017): 1307–1310; John Meurig Thomas, "Providing Sustainable Catalytic Solutions for a Rapidly Changing World: A Summary and Recommendations for Urgent Future Action," *Philosophical Transactions of the Royal Society* 376, no. 2110 (2018): 20170068.

CHAPTER 12: PETROLEUM FLIP-FLOP

1. D. L. Hoffmann et al., "U-Th Dating of Carbonate Crusts Reveals Neandertal Origin of Iberian Cave Art," *Science* 359, no. 6378 (2018): 912–915.
2. Miles Moore, "Exec: Expect Carbon Black Shortage in Five Years," *Rubber & Plastics News*, July 15, 2015, http://www.rubbernews.com/article/20150717/NEWS/307139999/exec-expect-carbon-black-shortage-in-five-years.
3. Steven C. Peterson and Nirmal Joshee, "Co-Milled Silica and Coppiced Wood Biochars Improve Elongation and Toughness in Styrene-Butadiene Elastomeric Composites While Replacing Carbon Black," *Journal of Elastomers & Plastics* (2018): 0095244317753653, http://journals.sagepub.com/doi/abs/10.1177/0095244317753653.
4. Hashim A. Alhashimi and Can B. Aktas, "Life Cycle Environmental and Economic Performance of Biochar Compared with Activated Carbon: A Meta-Analysis," *Resources, Conservation and Recycling* 118 (2017): 13–26.
5. 44–170 Mj/kg for the activated carbon compared with 1.1–16 Mj/kg for the biochar.
6. Emiliano Lepore, Francesco Bonaccorso, Matteo Bruna, Federico Bosia, Simone Taioli, Giovanni Garberoglio, Andrea Ferrari, and Nicola Maria Pugno, "Silk Reinforced with Graphene or Carbon Nanotubes Spun by Spiders," arXiv preprint arXiv:1504.06751 (2015), http://adsabs.harvard.edu/cgi-bin/bib_query?arXiv:1504.06751.
7. Kraig Biocraft Laboratories, http://www.kraiglabs.com.

CHAPTER 13: MY TESLA RUNS ON BANANA PEELS

1. W. Kolby Smith, Maosheng Zhao, and Steven W. Running, "Global Bioenergy Capacity as Constrained by Observed Biospheric Productivity Rates," *BioScience* 62, no. 10 (2012): 911–922, https://doi.org/10.1525/bio.2012.62.10.11.
2. Yifan Wang et al., "Converting Ni-Loaded Biochars into Supercapacitors: Implication on the Reuse of Exhausted Carbonaceous Sorbents," *Scientific Reports* 7 (2017): 41523.
3. Jiangfeng Li and Qingsheng Wu, "Activated Carbon Derived from Harmful Aquatic Plant for High Stable Supercapacitors," *Chemical Physics Letters* 691 (2018): 238–242.
4. Celia Hernández-Rentero et al., "Low-Cost Disordered Carbons for Li/S Batteries: A High-Performance Carbon with Dual Porosity Derived from Cherry Pits," *Nano Research* 11, no. 1 (2018): 89–100.
5. Marco Keiluweit et al., "Dynamic Molecular Structure of Plant Biomass-Derived Black Carbon (Biochar)," *Environmental Science & Technology* 44, no. 4 (2010): 1247–1253, https://pubs.acs.org/doi/abs/10.1021/es9031419.

6. Haoran Yuan et al., "Nonactivated and Activated Biochar Derived from Bananas as Alternative Cathode Catalyst in Microbial Fuel Cells," *The Scientific World Journal* (2014): 832850.

7. Xinda Li et al., "Flexible and Self-Healing Aqueous Supercapacitors for Low Temperature Applications: Polyampholyte Gel Electrolytes with Biochar Electrodes," *Scientific Reports* 7, no. 1 (2017): 1685.

8. Jung A. Lee et al., "An Electrochemical Impedance Biosensor with Aptamer-Modified Pyrolyzed Carbon Electrode for Label-Free Protein Detection," *Sensors and Actuators B: Chemical* 129, no. 1 (2008): 372–379.

9. Yong Yuan et al., "Applications of Biochar in Redox-Mediated Reactions," *Bioresource Technology* 246 (2017): 271–281.

CHAPTER 14: COOLING IT ON CARBON

1. Craig Somerton, *AC 2010-555: An Appropriate Technology Project: A Solar Powered Vaccine Refrigerator* (Washington, DC: American Society for Engineering Education, 2010).

2. Grant Banks, "World's Lightest Solid Material, Known as 'Frozen Smoke,' Gets Even Lighter," *New Atlas*, January 14, 2011, https://newatlas.com/worlds-lightest-solid-material-gets-even -lighter/17588.

3. Researchers show that in some tests, one type of biochar reduced nutrient concentrations and load in runoff while another type had an opposite effect. Once again, characterization of biochar types for particular applications matters. What is suitable for one use may not be suitable for another. Deborah A. Beck, Gwynn R. Johnson, and Graig A. Spolek, "Amending Greenroof Soil with Biochar to Affect Runoff Water Quantity and Quality," *Environmental Pollution* 159, no. 8–9 (2011): 2111–2118; Kirsi Kuoppamäki et al., "Biochar Amendment in the Green Roof Substrate Affects Runoff Quality and Quantity," *Ecological Engineering* 88 (2016): 1–9.

4. Ben Coxworth, "Paper Waste Converted into Eco-Friendly Aerogel," *New Atlas*, February 6, 2016, https://newatlas.com/paper-waste-cellulose-aerogel/41695.

5. Nick Lavars, "Wound-Plugging XStat Syringe Saves Its First Life on the Battlefield," *New Atlas*, May 24, 2016, https://newatlas.com/wound-xstat-bleeding-battlefield/43515.

CHAPTER 15: PETS, PERSONAL CARE, PILLOWS, PAINTS, AND PLASTERS

1. Roskill Market Reports, "Bentonite Global Industry, Markets & Outlook (2015)," https://roskill .com/market-report/bentonite; S. Sutton, "Raw & Manufactured Materials: 2016 Overview," *Ceramic Industry* 166:1 (January 2016).

2. "Curaprox Launches Its Black Is White and White Is Black Toothpastes in the United States," *DentistryIQ*, June 6, 2016, http://www.dentistryiq.com/articles/2016/06/curaprox-launches -its-black-is-white-and-white-is-black-toothpastes-in-the-united-states.html.

3. Amanda L. Dawson et al., "Turning Microplastics into Nanoplastics through Digestive Fragmenta- tion by Antarctic Krill," *Nature Communications* 9, no. 1 (2018): 1001.

4. Yanling Zheng et al., "Effects of Silver Nanoparticles on Nitrification and Associated Nitrous Oxide Production in Aquatic Environments," *Science Advances* 3, no. 8 (2017): e1603229.

5. Eoin McGillicuddy et al., "Charcoal as a Capture Material for Silver Nanoparticles in the Aquatic Environment," *EGU General Assembly Conference Abstracts* 19 (2017): 5023.

6. Mark Ainsworth and James Anthony Butcher, "Protective Clothing," US Patent 3,586,596, filed September 16, 1966, and issued June 22, 1971, https://patents.google.com/patent /US3586596.

7. "Bamboo Charcoal," Simplifi Fabric, https://www.simplififabric.com/pages/bamboo-charcoal.

8. "Activated Carbon Blanket," Nirvana Safe Haven, https://nontoxic.com/activatedcarbonblanket/activatedcarbonblanket.htm.
9. "Insomnia,"Mt.MeruPte.,https://www.mtmeru.com/bamboo-charcoal/our-products-insomnia.htm.
10. Hans-Peter Schmidt and Kelpie Wilson, "55 Uses of Biochar," *The Biochar Journal*, 2014.
11. Hiroshi Matsumoto, Midori Yokogoshi, and Yuki Nabeshima, "Effects of Moisture Controlled Charcoal on Indoor Thermal and Air Environments," *AIP Conference Proceedings* 1892, no. 1 (2017): 160003, https://aip.scitation.org/doi/10.1063/1.5005770.

CHAPTER 16: CARBON DETOX

1. Murray M. Finkelstein, "Silica, Silicosis, and Lung Cancer: A Risk Assessment," *American Journal of Industrial Medicine* 38, no. 1 (2000): 8–18.
2. Teresa Jane Brown et al., *World Mineral Production 2004–2008* (Nottingham, UK: British Geological Survey, 2010).
3. David L. Ashley et al., "Blood Concentrations of Volatile Organic Compounds in a Nonoccupationally Exposed US Population and in Groups with Suspected Exposure," *Clinical Chemistry* 40, no. 7 (1994): 1401–1404.
4. IARC Working Group, "Formaldehyde, 2-butoxyethanol and 1-tert-butoxypropan-2-ol," *IARC Monographs on the Evaluation of Carcinogenic Risks to Humans* 88 (2006): 1.
5. S. P. Newman, "Metered Dose Pressurized Aerosols and the Ozone Layer," *European Respiratory Journal* 3, no. 5 (1990): 495–497.
6. T. M. Perera et al., "Indoor Air Quality and Human Activities in Buildings," presentation, Civil Engineering Research Exchange Symposium, Matara, Sri Lanka, 2013.
7. Alison Cohen, Sarah Janseen, and Gina Solomon, *Clearing the Air: Hidden Hazards of Air Fresheners* (Natural Resources Defense Council, 2007).
8. Anne C. Steinemann et al., "Fragranced Consumer Products: Chemicals Emitted, Ingredients Unlisted," *Environmental Impact Assessment Review* 31, no. 3 (2011): 328–333.
9. Stanley M. Caress and Anne C. Steinemann, "Prevalence of Fragrance Sensitivity in the American Population," *Journal of Environmental Health* 71, no. 7 (2009).

CHAPTER 17: CRYPTOCARBON

1. Institute for Global Environment Strategies, 2018.
2. Amar Bhattacharya, Jeremy Oppenheim, and Nicholas Stern, "Driving Sustainable Development through Better Infrastructure: Key Elements of a Transformation Program," Brookings Global Working Paper Series, 2015.
3. Alex de Vries, "Bitcoin's Growing Energy Problem," *Joule* 2, no. 5 (2018): 801–805.
4. Chris Baraniuk, "Bitcoin Energy Use in Iceland Set to Overtake Homes, Says Local Firm," *BBC News*, February 12, 2018.
5. Eric Holthaus, "Move Over, Bitcoin Bros: A Green Cryptocurrency Is Here," *Grist*, August 1, 2018.
6. G. Booman, A. Craelius, B. Deriemaeker, G. Landua, W. Szal, and B. Weinberg, Regen Network Whitepaper Version 1.1 (September 7, 2018), http://regen-network.gitlab.io/whitepaper/White Paper.pdf.
7. S. Adams, "Clayton Christensen on What He Got Wrong about Disruptive Innovation," *Forbes*, October 2016.
8. Jonathan Shapiro, *Lawyers, Liars, and the Art of Storytelling: Using Stories to Advocate, Influence, and Persuade* (American Bar Association, 2016).
9. D. Messner, "A Social Contract for Low Carbon and Sustainable Development: Reflections on Non-Linear Dynamics of Social Realignments and Technological Innovations in Transformation Processes," *Technological Forecasting & Social Change*, 98 (2015): 260–270.

Notes

CHAPTER 18: THE COOL LAB

1. Seachar.com, "Estufa Finca Project," http://seachar.org/estufa-finca-project.
2. *Leucaena zeucocephala, Vigna unguiculata, Clitoria ternatea, Desmodium distortum, Psophocarpus tetragonolobus, Macroptilium lathyroides, Phaseolus calcaratus, Brassica napus*, and *Manihot esculenta*, for instance, based upon estimates for the Commonwealth from Maya Mountain Research Farm in Belize.
3. *Beyond the Tragedy of the Commons*, Stockholm Whiteboard Seminars, April 3, 2009, https://www.youtube.com/watch?v=ByXM47Ri1Kc.
4. David Sloan Wilson, Elinor Ostrom, and Michael E. Cox, "Generalizing the Core Design Principles for the Efficacy of Groups," *Journal of Economic Behavior & Organization* 90 (2013): S21–S32.
5. W. Kolby Smith, Maosheng Zhao, and Steven W. Running, "Global Bioenergy Capacity as Constrained by Observed Biospheric Productivity Rates," *BioScience* 62, no. 10 (2012): 911–922, https://doi.org/10.1525/bio.2012.62.10.11.
6. Frank Michael and Albert Bates, "Optimized Potentials for Soil Sequestration of Atmospheric Carbon," in *Biomass Energy Crops,* Maletta et al., eds. (London: Baker and Taylor 2018, in press).
7. C. Werner, H.-P. Schmidt, D. Gerten, W. Lucht, and C. Kammann, "Biogeochemical Potential of Biomass Pyrolysis Systems for Limiting Global Warming to 1.5°C," *Environmental Research Letters* 13, no. 4 (2018): 044036.
8. Smith et al., note 5.

CHAPTER 19: GRASSROOTS SOLUTIONS

1. World Economic Forum, "The Green Investment Report: The Ways and Means to Unlock Private Finance for Green Growth," 2015, http://www3.weforum.org/docs/WEF_GreenInvestment_Report_2013.pdf.

CHAPTER 20: CIVILIZATION 2.0

1. Charles C. Mann, *The Wizard and the Prophet: Two Remarkable Scientists and Their Dueling Visions to Shape Tomorrow's World*, 1st ed. (NY: Knopf, 2018).
2. David Fleming, *Lean Logic: A Dictionary for the Future and How to Survive It* (White River Junction, VT: Chelsea Green, 2016).
3. Timothy J. Garrett, "No Way Out? The Double-Bind in Seeking Global Prosperity Alongside Mitigated Climate Change," *Earth System Dynamics* 3, no. 1 (2012): 1–17.
4. Thich Nhat Hanh, *The Heart of Understanding: Commentaries on the Prajñaparamita Sutra* (Berkeley: Parallax Press, 1988), 3–5, http://www.parallax.org.
5. Daniel Wahl, *Designing Regenerative Cultures* (Triarchy Press 2016).
6. Adm. Hyman Rickover, testimony before Congress, published in AEC, *Authorizing Legislation: Hearings Before the Joint Committee on Atomic Energy* (1970), 1702.
7. Each gigaton of carbon removed subtracts 0.47 ppm from global atmospheric concentration (IPCC Working Group 1). However, the ocean is in approximate equilibrium with the atmosphere, so decreasing the atmospheric concentration actually requires about double the carbon removal to achieve a desired net effect. Removing 325 GtC net from the atmosphere requires gross removal of 650 GtC, which could return the atmosphere to preindustrial concentrations of 260 ppm. That said, because forced reduction of atmospheric CO_2 has no historical precedent, there are many unknowns.

INDEX

Index

Index

Index

Index

ABOUT THE AUTHORS

Albert Bates is one of the founders of the intentional community and ecovillage movements. A lawyer, scientist, and teacher, he has taught village design, appropriate technology, and permaculture to students from more than sixty countries. His books include *Climate in Crisis* (1990); *The Post-Petroleum Survival Guide and Cookbook* (2006); *The Biochar Solution* (2010); and *The Paris Agreement* (2015).

Kathleen Draper has been deeply involved in many areas of biochar research, communication, and outreach. She routinely collaborates with biochar experts from around the globe as a board member of the International Biochar Initiative (IBI), a moderator for IBI's biochar education webinar series, and the US director of the Ithaka Institute for Carbon Intelligence. She has lectured on biochar in several countries and provides consulting services to companies entering the biochar industry. She is editor of the online review *The Biochar Journal*.